HORSEKEEPING SKILLS LIBRARY

Stablekeeping

A VISUAL GUIDE TO SAFE AND HEALTHY HORSEKEEPING

CHERRY HILL

PHOTOGRAPHY BY
RICHARD KLIMESH

Storey Publishing

The mission of Storey Publishing
is to serve our customers by publishing practical information
that encourages personal independence in harmony with the environment.

Edited by Deborah Burns and Marie Salter
Cover design by Eugenie Delaney
Cover photograph by Richard Klimesh
Text design and production by Susan Bernier
Production assistance by Jennifer Jepson Smith
Photographs by Richard Klimesh
Illustrations on pages 12, 17, 39, 42–46, 48, 55, 72, 78, 82, 84, 104, 106, 111, 133 by
    Alison Kolesar; all others by Richard Klimesh
Indexed by Susan Olason/Indexes & Knowledge Maps

The information in this book is true and complete to the best of our knowledge. All recommendations are made without guarantee on the part of the author or Storey Publishing. The author and publisher disclaim any liability in connection with the use of this information. For additional information please contact Storey Publishing, 210 MASS MoCA Way, North Adams, MA 01247.

Storey books are available for special premium and promotional uses and for customized editions. For further information, please call 1-800-793-9396.

Printed in the United States by Versa Press
20 19 18 17 16 15 14 13 12 11 10 9

Library of Congress Cataloging-in-Publication Data

Hill, Cherry, 1947–
    Stablekeeping : a visual guide to safe and healthy horsekeeping / Cherry Hill.
        p. cm.
    Includes index.
    ISBN-13: 978-1-58017-175-5
    ISBN-10: 1-58017-175-3
    1. Stables—Management. 2. Horses. I. Title.

SF285.35 .H56 2000
636.08'3 21–dc21

99-045164

Contents

Preface

Stablekeeping is the coordination of day-to-day tasks involved in keeping a barn running smoothly. Whether you're caring for one horse or fifty, whether you have your own barn or are working in someone's stable, the ideas in this book can help improve the safety, efficiency, and atmosphere of your stable environment. Ultimately, good stablekeeping will minimize repair, feed, and veterinary bills and provide a safe and healthy living environment for horses as well as a pleasant workplace for the horses' people.

When it came time for me to choose the equine "cover girl" for this book, it was an easy choice. Sassy Eclipse, a 25-years-young Quarter Horse mare, is often mistaken even by experienced horsemen to be half that age! But here she is after more than 20 years of riding and six foals, looking just as bright and "sassy" as the day I bought her as a yearling. Since she was born, Sassy has had the good care that is outlined in this and my other horsekeeping skills books. By paying special attention to parasite control, sanitation, proper nutrition, and safety, Sassy and I plan to enjoy many more years together.

Acknowledgments

This stablekeeping book would not be complete without introducing my longtime husband and horsekeeping partner, Richard Klimesh. Not only is he a master at designing, building, and maintaining healthy and safe horse facilities, but he is my able partner in the day-to-day chores of caring for our horses. Our teamwork is energizing. Together we choreograph our daily chore routines and special tasks such as unloading hay and spreading manure. The hours are long and the work is sometimes hard and dirty, but it gives us a great deal of satisfaction and joy to pause in our quiet barn and just listen to the horses munching their hay.

Thanks to the following for help with the photos for this book:

Mrs. Putnum Davis, Staatsburg, New York

Farnam Companies, Inc., Phoenix, Arizona

Hamilton Products, Ocala, Florida

Horseware Triple Crown Blanket, Kinston, North Carolina

Ron Johnson, Johnson Barns and Trailers, Phoenix, Arizona

Detlef Juerss, Steinbau Construction, Pleasant Valley, New York

Kalglo Electronics Co., Inc., Bethlehem, Pennsylvania

Loveland Industries, Greeley, Colorado

Spalding Laboratories Biological Fly Control, Arroyo Grande, California

Wilsun Equestrian, Alpharetta, Georgia

✦ BARN FEATURES ✦

The atmosphere and safety of all activities within a barn are determined by three main barn features: the brightness and placement of lights, the amount of fresh air, and the material and condition of the aisle floor. These features can make the difference between a place you *have* to go to do chores, and a place you *want* to go to spend time with your horses. The image of a person stumbling over a wavy dirt floor in a dimly lit stable while breathing foul air might be suited to a Dickens novel, but this is not the stuff of good stablekeeping. No matter what size or shape a stable is, or whether it's old or new, the environment can often be improved by adding or moving lights, installing windows and vents, and remodeling floors.

Lighting

General, or ambient, lighting lets you see everything in the stable and enables you to move around safely without falling over things. Natural lighting can be supplied by a combination of overhead lights, skylights, windows, translucent wall panels, and open doors. Light-colored walls and ceilings reflect light and make the most of ambient light.

All lightbulbs in a stable should be protected from dust and impact. Dust on the surface of a bulb reduces light output and can be a fire hazard. Lightbulbs break easily upon impact and can cause injury from shards of glass and fire from the hot filament.

Standard incandescent lightbulbs are cheap to buy and easy to replace, but when compared to fluorescent and halogen lights, they are inefficient and burn out quickly. An incandescent bulb's limited light output is further restricted by the thick glass cover and metal grille needed to protect it. Halogen lights are very bright and usually come encased in sturdy, compact, dustproof fixtures that are easy to install. Fluorescent bulbs are relatively cool, give off more than twice the light of an incandescent, and are cheap to operate, but the most common fixtures are 4 feet (1.2 m) long, which means they take up a lot of space and catch a lot of dust in a barn environment. Fluorescent bulbs should be protected from dust, moisture, and impact by a transparent plastic cover.

Task lighting, or work lighting, is for small areas where you need to see what you're doing without shadows and glare. In the feed room, for example, you need to weigh grain, read labels, and measure feed and supplements accurately. The amount of light you need depends on the task, the room, and the location of the light. In some feed rooms a single 100-watt incandescent bulb might be enough, while others might require several bulbs in different locations. Locate task lighting so it doesn't glare in your eyes as you do your work. Also, install plenty of plug-in outlets throughout the stable, for using temporary lights where they're needed, such as for the veterinarian or farrier.

Outside lights by entrances can make your trip to the stable at night safer. Lights over pens can be invaluable when you need to catch or doctor a horse after dark.

Lighting

1.1 FIBERGLASS PANELS

One of the easiest ways to get natural light into a barn is with domed skylights or with less expensive translucent fiberglass panels. Anytime openings are put in a roof, there's a chance of leaking, so roof skylights must be installed properly. This stable has a solid roof but has fiberglass panels on the gable ends and on the upper part of the walls and sliding doors to provide ample ambient light during daylight hours.

1.2 JELLY JAR

A "jelly jar" protects this incandescent bulb from dust and moisture, while the heavy-duty cast aluminum guard shields the bulb from impact.

1.3 HALOGEN FIXTURE

This 150-watt halogen light provides plenty of illumination for a 12-foot by 12-foot (3.7-m by 3.7-m) stall. Sturdy halogen fixtures like this one with a removable glass cover are readily available.

1.4 INCANDESCENT FLOODLIGHTS

This adjustable fixture holds a pair of incandescent floodlights used to illuminate the exterior of a stable entrance.

1.5 HID YARD LIGHT

HID lights used as yard lights are commonly controlled by a photosensor that turns the light on automatically at dusk.

TYPES OF LIGHTING

TYPE OF BULB	DESCRIPTION	COMMENTS
Incandescent	Standard lightbulb; tungsten filament; classic "bulb" shapes	Cheapest to buy, quickest to go; only lasts 500–1,000 hours
Halogen or quartz	Tungsten filament inside a clear quartz tube filled with halogen gas; small tubes	Brighter, more efficient, and lasts longer than incandescent, but costs more; oil from fingers can cause the bulb to fail more quickly; very hot, keep away from combustible materials; lasts up to 4,000 hours
Fluorescent	No filament; electrodes charge mercury vapor, which causes phosphor coating to glow; commonly long tubes, but many other shapes	Coolest, most energy-efficient type of light; a 40-watt fluorescent bulb puts out twice the light of a 100-watt incandescent bulb, uses less than half the electricity, and lasts 10 times longer; some rated as long as 20,000 hours
High-intensity discharge (HID)	Filled with sodium, metal halides, or mercury vapor; electricity excites the gas to produce light; bulb shape	Requires special ballasts and fixtures; needs warm-up time; used mainly outdoors for streetlights and security

Ventilation

Ventilation can replace stale air, which contains ammonia and other products of waste decomposition, with fresh air, which is vital for a horse's and person's health. The more fresh air in a stable the better, as long as horses are not subjected to cold drafts. Increased airflow throughout the stable also minimizes heat buildup in the summer and accumulation of dust. If a stable smells bad or stuffy, there is almost surely inadequate ventilation. Ventilation is provided by windows, doors, roof vents, and exhaust fans.

1.6 STALL WINDOWS

These individual stall windows open to the outside and are protected by steel grilles on the inside of the window openings. Large sliding doors at the ends of a stable are an excellent way to provide varying degrees of cross-ventilation and light. Entrance doors should be a minimum of 6 feet (1.8 m) wide for safe leading of horses and easy passage of small carts for feeding and cleaning stalls. For vehicle access and easy delivery of hay and bedding, the doors should be a minimum of 12 feet (3.7 m) wide.

1.7 FANS

In stables with low ceilings, air can be forcibly moved by the use of exhaust fans placed high in the walls of the stalls and protected by sturdy metal bars. In addition, portable fans can be mounted in the aisles to move air through the stalls, which will cool horses and help dry the bedding.

Aisle Flooring

Because the main aisle receives a lot of traffic, the floor should be durable and easy to clean. To be safe, the floor should be relatively smooth and level so a horse or handler doesn't trip, yet have sufficient traction to prevent slipping. The softer and more shock-absorbing a floor is, the less fatiguing it will be to walk and stand on. Harder floors increase the likelihood of a person or horse being injured from falling.

The appearance of a floor, from high-tech to rustic, and the sound of a floor, from silent to clip-clop, will have a big influence on the character of a stable.

To decide which material is best for your stable, balance personal preferences with cost, availability, maintenance, and durability. The accompanying chart compares various materials that have been used for aisle floors in barns.

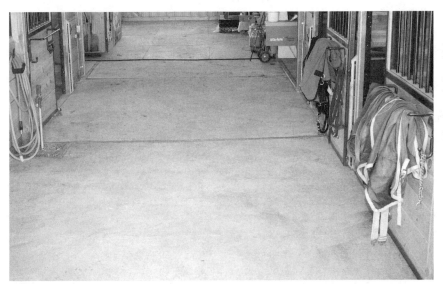

1.8 CONCRETE AISLE
A concrete aisle, when properly textured, has good traction and is easy to clean. Expansion joints built into the floor at 10-foot (3-m) intervals help prevent the floor from cracking.

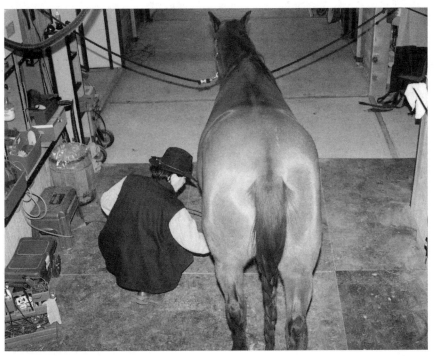

1.9 RUBBER MATS
Rubber mats over concrete produce a comfortable grooming-area floor that has cushion and reduces fatigue for both horse and groom.

TYPES OF FLOORING

MATERIAL	DESCRIPTION	ADVANTAGES
Dirt	Can vary from loose and sandy to hard clay	Inexpensive; already in place or easy to install; great traction; very quiet; can have good shock-absorbency
Wood shavings, chips, shredded bark	Bagged or bulk	Inexpensive; soft and quiet; looks good if kept clean and raked; smells good when fresh; helps control mud by soaking up moisture; can reduce dust from the dirt base
Sand	A very fine form of crushed rock, varies from very clean and almost dust-free "masonry sand," to dirty and dusty "fill sand"	Inexpensive; very quiet and soft
Road base	A mixture of dirt and gravel; varies in composition depending on location	Makes a good base for almost any kind of solid flooring; good choice to place on top of dirt for a temporary floor
Gravel	Crushed rock that is coarser than sand; various sizes and purities	Drains extremely well; makes an excellent base for many types of solid flooring
Wood	2" (5.1-cm) planks laid across 4 x 4s set directly on a draining base, such as road base, limestone, or gravel	Provides a unique feeling and sound; one of the warmest types of floors; can be taken up and reinstalled relatively easily
Brick	Small, rectangular blocks, 2¼" x 3¾" x 8" (5.7 cm x 9.5 cm x 20.3 cm), usually made of clay and sand; types vary in hardness and durability; used as paving material for more than 5,000 years; commonly laid "loose," without mortar, in a bed of sand	If properly installed, will stay level and last indefinitely; common unglazed paving brick has good traction; like wood, adds a unique feeling and sound to a stable
Rubber brick	A new take on an ancient theme; from 1⅝" to 2" (4.1 to 5.1 cm) thick, and in a variety of shapes and colors	Durable; very quiet and resilient; excellent traction wet and dry; easier to fit along edges than clay brick, since it can be cut with a knife or band saw
Rubber tile	Like rubber brick, available in a variety of colors; 1" to 1¼" (2.5 to 3.2 cm) thick; designed to go over a solid, impermeable surface such as concrete or asphalt	Same advantages of rubber brick
Concrete	A very hard, extremely durable material composed of cement, sand, gravel, and water	Properly installed, is permanent and maintenance-free; easy to clean if texture is not too deep
Asphalt	A dull black paving material, usually a combination of asphalt, sand, and powdered limestone; applied hot onto a compacted base, leveled, and then smoothed by a heavy roller — not a do-it-yourself deal	Cheaper than concrete
Rubber mats	From ½" to ⅞" (12.7 to 22 mm) thick, usually in 4' x 6' (1.2-m by 1.8-m) pieces; applied over concrete or tamped gravel or road base	Quiet; good cushion; easy to clean; easy to install; very good traction when dry and sufficient when wet; can readily be used on just a portion, such as the grooming area

DISADVANTAGES	COMMENTS
Dusty when dry; muddy when wet; moisture and traffic cause "war zone" effect of craters and humps; high maintenance	Logical for a temporary floor; poor in the long run
A serious fire hazard; loose surface is a black hole for items such as hoof picks, combs, and horseshoe nails; breeze can blow shavings everywhere; if not dampened, shavings can add to air pollution	All things considered, this step in floor evolution should be skipped entirely
Messy; dusty; shifts under foot and is easily thrown to the rafters by a pawing, fractious, or rambunctious horse	If your native soil is sand, mix it with dirt or road base and cover it with solid flooring
Same drawbacks as dirt	Good temporary floor; unsuitable as a permanent floor
Like sand, unstable and usually dusty; sharp gravel is hard on boots and hooves; round gravel rolls and is unstable	Unsuitable as a finished floor
Slippery when wet; shoes with traction devices can splinter wood and "stick" to it, possibly causing joint injuries; boards of different thicknesses make an uneven surface; absorbs urine and water; difficult to keep clean	Okay; can provide an "old-time" atmosphere for many years
Can settle unevenly and develop waves (which may or may not add desired ambience to the stable); difficult to sweep clean because of the joints between the bricks	Good; unique *clip-clop* sound and appearance
Very expensive; surface texture makes it difficult to sweep	Excellent; safe and durable; classy-looking
Very expensive; doesn't provide quite as much cushion as rubber brick; moisture can accumulate between tile and impermeable base	Good; safe, durable
Smooth finish is very slippery, especially for shod horses and when wet; very difficult to modify or replace; surface most likely to cause injury from impact; can be noisy with shod horses	Very good permanent floor
Especially slippery when wet and for shod horses; not as durable as concrete; can heave and crack during seasonal freeze-thaw cycles; hardness changes with temperature, from soft and tacky on hot days to hard and slick when cold	Unsuitable for a stable, better for a driveway
Can bulge at seams with moisture and temperature fluctuations; interlocking mats have puzzle-piece edges that prevent this; *Caution:* old conveyor belting is dangerously slippery when wet	Excellent; best over solid base such as concrete, never over sand; a surface pattern on mats does not significantly improve traction

✦ STALLS ✦

A stall is a horse's indoor living quarters, and often the main reason a barn exists. Stalls are best reserved for tending to sick horses, for sheltering horses from extreme weather, and for keeping horses clean if they are ridden or exercised daily. Generally, the more time a horse spends outside in fresh air and sunshine, the healthier he will be.

A box stall or "loose" stall is safer, healthier, and more comfortable than a "tie" stall, which is a narrow chute where a horse stands tied. A box stall should be large enough to allow a loose horse to move about freely and to lie down and get up without banging into the walls. It should be roomy enough to accommodate separate areas for a horse's three main activities: eating, lying down, and defecating/urinating. Given enough space, a horse will often defecate in one spot in a stall. A stall that is too small will cause a horse to walk through his manure, churning it into the bedding and mixing it with his hay. This wastes bedding *and* feed, and makes cleaning the stall much more labor intensive. A stall that is too large, on the other hand, requires much more bedding to prepare and takes more time to clean.

The more time a horse spends in a stall, the roomier the stall should be. For example, a 10-foot by 12-foot (3-m by 3.7-m) stall could be adequate for a 1,000-pound (454-kg) riding horse that's in a stall for only a few hours to cool out after a ride. If the horse spends most of the time in the stall, however, 12 feet by 14 feet to 12 feet by 20 feet (3.7 m by 4.3 m to 3.7 m by 6.1 m) would be more appropriate.

The dividing wall between two stalls should be high enough so horses cannot reach one another over the top. Feeders should not be on the dividing wall unless the wall is solid where the feeders are, to prevent horses from seeing one another when they are eating, which can lead to aggressive behavior.

STALL SIZES

STALL SIZE		SUITABLE FOR
Feet	**Meters**	
10 x 10	3 x 3	**Pony:** Under 14.2 hands and usually under 800 pounds (363 kg)
12 x 12	3.7 x 3.7	**Standard horse:** Up to 16 hands and 800 to 1,100 pounds (363–499 kg)
12 x 14	3.7 x 4.3	**Large horse** (Warmblood): More than 16 hands and 1,100 to 1,400 pounds (499–635 kg)
12 x 16	3.7 x 4.9	**Draft horse:** More than 16 hands and over 1,400 pounds (635 kg)
12 x 18	3.7 x 5.5	**Foaling stall:** Up to 1,400-pound mare with foal (635 kg)

Kick Walls

Horses kick when they're underexercised, playful, bored, irritated, or uncomfortable. When they kick the stall walls, they can easily knock the barn siding off or kick through the siding, resulting in serious leg injuries. Horses can also injure a horse next door if the stall dividers are not strong enough. The walls of a stall should be lined up to 4 feet (1.2 m) from the floor with a material strong enough to withstand a horse's kick. Some materials commonly used are 1½-inch-thick (3.8-cm) tongue-and-groove (T&G) lumber, full thickness (2 inches [5.1 cm]) rough sawn lumber, and ¾-inch (19.1-mm) plywood covered with sheet metal. Rubber mats that are ⅝-inch (15.7-mm) thick make an excellent lining for stalls. They cushion the blow of a kick, thus protecting the horse's legs and feet. Also, mats don't mar or dent like wood or metal surfaces do.

Grilles

Metal bars, or grilles, on the upper portion of stall walls allow containment of a horse with minimal ventilation interference. Grilles also allow a horse to see out, which helps prevent boredom, and allow a manager to see into the stall without opening the door. Spaces between the bars that are large enough for a horse to put his head through will almost surely lead to injury because horses are easily startled and move quickly when they are surprised. The safest bars are spaced so a horse can't get a hoof through them if he rears up.

Instead of a grille, it is sometimes better to have a solid partition at least 7 feet (2.1 m) high between stalls to prevent horses from seeing and smelling one another. Playing often leads to fighting and injury.

Heaters

A heater is used primarily to help a hot horse cool out gradually, to prevent a wet horse from being chilled as he dries, and to provide warmth to a sick or injured horse. It's important that the heat source is located so a horse can move away from the heat if he gets too warm.

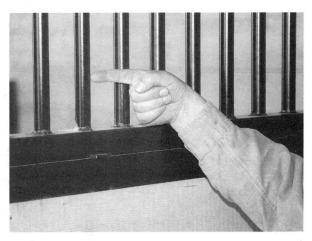

2.1 A 2-inch (5.1-cm) space, seen here as the length of two finger joints, is very safe. A 4-inch (10.2-cm) space between bars is considered the maximum safe grille spacing for riding horses (800–1,100 pounds [363–499 kg]).

infrared heater

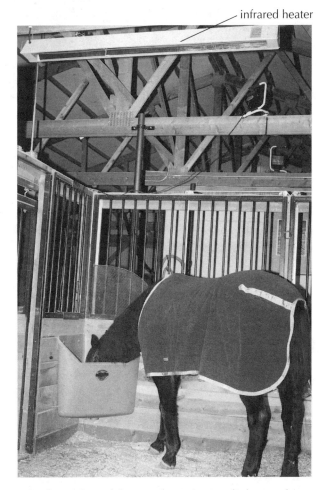

2.2 An infrared heater (see Appendix) or a heat lamp heats objects, not air. This is much healthier for a horse than heating the entire barn.

Doors

2.3 STALL DOOR WIDTH

The opening of a stall door should be 4 feet (1.2 m) or wider and 8 feet (2.4 m) high. This allows you to lead a horse through without rubbing against the sides or the horse bumping his head, and to get a cart through the doorway without banging the wheels' hubs. The edges of door openings should be rounded, such as with a farrier's rasp, a wood plane, or an electric router, to prevent injury. One or two unpleasant collisions with a doorway can make a horse "door shy": He'll either balk at the doorway or tend to rush through.

A

B

2.4 DOOR ROLLER AND STOP

Sliding doors require very little space to open and close and are convenient to operate. When the door is closed, make sure the bottom of the door is secured at one end with rollers (A) and at the other end with an L-shaped stop (B) to prevent the horse from pushing the door out and getting a leg caught between the door and the frame.

2.5 DROP-DOWN WINDOW

Horses are social animals; many will be less bored and more content if they can see other horses and keep an eye on the barn activities from their stall. A window in the stall that opens to the aisle allows a horse to put its head out and see what's happening. A drop-down window has an advantage over a swinging window in that it doesn't need to be fastened open. One drawback to letting a horse put his head out of the stall is that he is then in a perfect position to bang the wall with his front hoof and knee. Also, horses can develop a dangerous habit of lunging and biting at human and equine passersby, especially if the barn aisle is narrow.

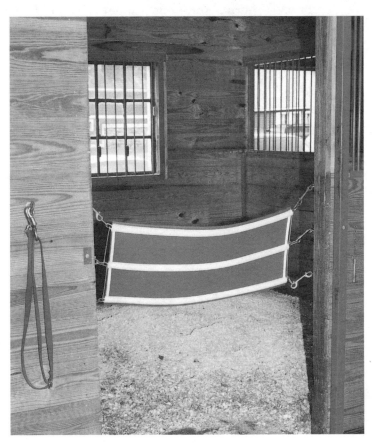

2.6 DOOR GUARD

A rope, chain, or strap door guard gives a horse more freedom than a window and may make him feel less confined. With an open doorway, however, bedding will find its way into the aisle. Some horses will try to push through a door guard or will lunge at persons or horses passing by. It's best to unhook the guard to enter a stall, because when ducking under, a person is in a vulnerable position to be kicked.

Doors (continued)

2.7 DOOR SILL

A 2 x 6 door sill with metal antichew strips has been added to this stall doorway to help keep bedding from littering the aisle. This simple door guard, consisting of a length of chain covered with rubber tubing, would not be substantial enough to keep some horses in.

2.8 "IN AND OUT"

A pen adjoining the stall provides a convenient means to turn a horse out without having to halter him. Often, however, a horse that's allowed to move freely between the stall and the outside will develop the irritating habit of entering the stall only to defecate and urinate. To minimize stall cleaning and bedding use, latch the stall door closed when the horse is let into the pen. Protection from sun and rain can be provided by a roof overhang.

DUTCH DOORS

A Dutch door (one that is split into top and bottom sections) is very useful between stalls and adjoining pens. The top can be fixed open to allow a horse to look out. The bottom door can be opened while the top remains closed, so a person can duck through for feeding, for example, without the horse following. This arrangement is also useful for making a "creep" area, which allows the foal to enter the stall to eat and rest in the shade without the mare following. (*Note:* The design of some split doors prevents the bottom portion from being opened separately.)

Latches

Horses can be surprisingly dexterous with their lips, and some seem to enjoy the challenge of opening latches. A horse-proof latch is constructed in such a way that a horse cannot undo it. Any latch is horse-proof if a horse can't reach it. Many latches can be made horse-proof by the addition of a snap.

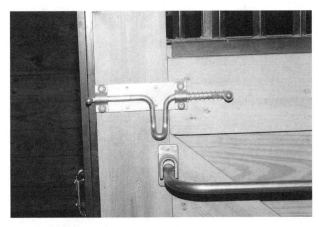

2.9 SPRING BOLT LATCH
This type of sliding bolt latch is commonly used on both hinged and sliding doors. This latch slides to the left and anchors in a keeper attached to the stall wall. When the latch is open, the sliding bolt must retract past the edge of the door (as shown here). If it sticks out, it could catch on a horse or person, causing injury or damage to clothing.

2.10 SLIDING BOLT LATCH
The handle of this type of sliding-bolt latch turns downward when closed to prevent a horse from opening it.

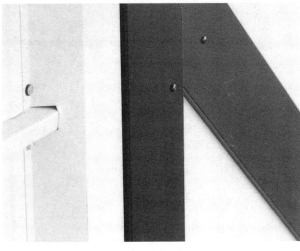

2.11 TURNING LATCH
This barn door–style turning latch is commonly used on outer stall doors that swing on hinges. One problem with this latch is that it protrudes about 1½ inches (3.8 cm), and some horses would love to rub on the hard edges.

2.12 LOWER LATCH
As with a sliding door, a lower latch on this hinged door prevents a horse from pushing the bottom of the door out and getting a leg caught between the door and the jamb. It also prevents the door from twisting and stressing the hinges if a horse kicks, paws, or rubs on the door.

Stall Flooring

A good stall floor greatly enhances the efficiency of stable management and saves time and money in the long run. In most cases, the use of stall flooring reduces the amount of bedding used when compared to dirt floors (see chapter 9). A poor stall floor can be a management nightmare. Primarily, a stall floor should be comfortable for the horse, safe, and easy to clean. Also, stall flooring should be fairly easy to install, affordable, and able to withstand abuse from hooves, shoes, and the acids in manure and urine. A stall floor must be safe when wet so a person or horse won't slip and be injured. A foaling mare, especially, needs good traction when her water breaks and she tries to lie down or get up. Also, newborn foals have difficulty standing up on some mats made slippery by placental fluids.

Natural soil is the cheapest and easiest floor to install because it requires no extra material; sand is not recommended, as it can be ingested by the horse and cause colic. A level dirt floor can quickly become an uneven mess from the urinating, pawing, and habitual movements of a horse. Concrete, on the other extreme, will maintain a level surface for years, but the hard surface is very stressful to a horse's legs and feet, and much bedding must be used to make concrete comfortable enough for a horse to lie on.

A good stall flooring, when properly installed on well-drained soil, will require minimal maintenance and eliminate the need to replace the dirt base, as is customary on an annual or semi-annual basis with a dirt stall. Flooring eliminates the hills and valleys, cuts bedding requirements to about half, and makes stalls easier and quicker to clean and disinfect.

STALL FLOORING

MATERIAL	COST PER SQ. FT.	BENEFITS	DRAWBACKS
Soil (clay, dirt)	$0.08	Good traction and cushion; inexpensive	High maintenance; difficult to clean; can be dusty
Gravel, sand	$0.08	Good traction and cushion; inexpensive	High maintenance; difficult to clean; can be dusty; danger of colic
Asphalt (blacktop)	$2.40	Easy to clean; low maintenance	Cold, hard, and abrasive; slippery; hard on legs and feet
Concrete	$3.00	Easy to clean; good traction if textured; maintenance-free	Cold, hard, and abrasive; hard on legs and feet; slippery if not textured
Wood	$1.50	Nice sound; warm	Slippery when wet; difficult to clean and disinfect
Draining flooring	$2.50 to $3.50	Minimal bedding required; moisture doesn't pool but drains through	Urine accumulates under flooring
Conveyor belting	$0	Free in some areas; easy to clean	Very slippery when wet; difficult to handle and install
Rubber mats	$1.60 to $2.40 (plus cost of base)	Quiet; good cushion and traction; easy to clean	Can be difficult to handle and install properly; can gap and buckle

Note: Pricing information is for general reference only.

There are generally two types of stall flooring: draining and nondraining.

Draining Flooring

A draining flooring allows urine to pass through holes in the material. Only the bedding immediately around the urine stream becomes wet and needs to be removed. One problem is that urine soaks through and accumulates in the gravel and dirt below. This can lead to odor problems, necessitating labor-intensive replacement of the base.

Draining flooring is only appropriate for barns on excavated, well-prepared sites or in locales with naturally well-draining soils. Using it on a site with poorly drained soils will quickly result in a soggy mess under the flooring, which is odorous and unhealthy.

Nondraining Flooring

A nondraining or solid flooring is made of an impermeable material that keeps moisture on top of the flooring where it's soaked up by the bedding. More bedding tends to be used with stall mats than with draining flooring, but it's still less than with a dirt floor, and less urine ends up under the stall floor. For a barn on poorly drained soils, it is best to consider solid flooring and plan to clean stalls more often, removing all wet bedding as soon as possible.

Advantages of solid stall mats include less dust, less ammonia and urine odor, better hoof health (e.g., lower incidence of thrush because the stalls stay drier), less hoof wear, more comfort, and a level surface that's easy to clean. A problem with solid mats is widening and buckling of the joints where mats meet. Bedding particles and feed chaff pack into gaps between mats and form a mortarlike joint. Fluctuations in moisture and temperature cause the mats and the packed chaff to expand and contract, creating wide gaps during the winter and bulging joints during warm weather. Fitting mats tightly against the walls of the stall when installing them can minimize this problem.

2.13 A wood floor is warm, has a romantic sound when horses walk on it, and falls in the middle of the flooring price range. It can be very slippery when wet, however, and the uneven edges and porous surface make it difficult to clean.

2.14 Two types of nondraining flooring and three types of draining flooring, left to right: used conveyor belting, ⅝-inch-thick (15.7-mm) rubber mat, one-piece perforated plastic mat, 1-foot (30.5-cm) square interlocking plastic tiles, tongue-and-groove gridwork tile.

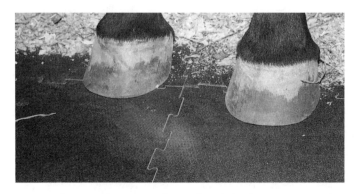

2.15 The puzzle-piece edges of interlocking mats make very tight joints that prevent the mats from separating. This eliminates the need to fit the mats snugly against the stall walls to maintain tight joints, as is necessary with noninterlocking mats.

Stall Mat Installation

The single most important factor that determines the successful performance of stall mats is floor preparation prior to installation. Stall mats should be put down over a firm, level gravel base. Sand should never be used under mats; it doesn't compact evenly and the mats will not stay level over shifting sand. If the mats are going on top of concrete or asphalt, the floor should be sloped to a drain to minimize the pooling of moisture beneath the mats.

1. Prepare the stall by adding 4 inches (10.2 cm) of clean gravel (⅜ inch [9.7 mm] or smaller) or stone dust (fines produced when making gravel).

2. Level the gravel using an 8-foot (2.4-m) 2 x 4 or similar tool. Place the board on edge and work it in a sawing motion while dragging it across the gravel in several directions, moving gravel from the high spots to fill in the low areas.

3. Pack the gravel with a tamping bar, and then level it again. Tamping bars are available in hardware and building supply stores. You can make one by welding a ¼-inch-thick (6.4-mm) steel plate, 5 to 8 inches (12.7–20.3 cm) wide, onto the end of a 6-foot (1.8-m) pipe, 2 inches (5.1 cm) in diameter. Any similar tool, a lawn roller, or even stomping feet can work to pack the base.

4. Once the stall base is prepared, you'll need a tape measure, a piece of chalk (a builder's chalk line would also work), a steel straightedge, and a standard utility knife for the installation.

5. Factory edges on mats are usually straight, so install the mats with factory edges against one another toward the center of the stall and your cut edges against the wall. The corners of the mats are the weakest part of the floor. Whenever possible, stagger the mats so four corners do not come together in one place. This minimizes the chances of the corners bulging up as bedding works into the joints (see page 97).

INSTALLATION TIP

Rubber mats are relatively easy to cut with a sharp utility knife, but not in one pass. Apply only moderate pressure to the knife and make several strokes to cut through the mat. With the mat on fairly level ground, place a 2 x 4 under the mat beneath the line you're cutting so the mat naturally bends down on either side. This opens the cut as you go and prevents the mats from pinching the knife blade. Blades are cheap, and you'll be surprised at how much better a sharp blade cuts than a dull one. Using a power saw is not a good idea; the friction melts the rubber, which grabs the blade and strains the saw motor. This method also produces great quantities of unhealthy smoke and harmful rubber dust.

3

✦ THE TACK ROOM ✦

One thing is certain about a tack room — it can never be too big. As the hub of the barn, the tack room serves as a climate-controlled tack storage vault, a records center, a tack repair shop, a closet, a laundry, and a pharmacy. Windows are often omitted from a tack room because they take up valuable wall space that could be used to store tack, and because sunlight can be harmful to leather tack. Also, a thief could gain entrance through a window, so if your tack room does have windows, protect them with secure iron grillwork.

Many items kept in the tack room are sensitive to extremes in temperature and humidity. High humidity is especially hard on leather tack, causing it to mold and mildew, while low humidity tends to dry leather out, causing it to crack. Depending on your climate, a humidifier or dehumidifier can be used to maintain humidity within the optimum range of 40 to 50 percent. Temperature extremes aren't as hard on tack items, but they can harm horse-care products such as medications, lotions, and creams.

Insulating the walls, floor, and ceiling of a tack room, and making sure doors and windows seal tightly, will make controlling humidity and temperature easier. This will also minimize dust and help keep out rodents and insects. A tightly sealed room can get stuffy, however, so some means of ventilation, such as screen doors and/or operable vents in the walls and ceiling, should be provided.

Tack Room Details

A tack room should be designed to fit your needs. The one below features a bridle rack with a shelf above it for extra saddle blankets, coolers, pads, leg wraps, and other supplies. Beneath the bridles, three fold-down English saddle racks are mounted to the wall. Tack hooks hanging from the ceiling are used to store ropes, bridles, and other gear waiting to be cleaned or repaired. The moveable Western saddle stand preserves the saddle's shape and makes a great cleaning stand for any saddle. Blanket rods are used for storing seasonal blankets and sheets and for allowing saddle blankets and pads to dry after use. The corner shelves that hold many small grooming and veterinary items are designed to use otherwise wasted corner space. The small refrigerator for medicines and beverages (see photo 3.8) sits on a file cabinet that contains warranties, registration papers, and veterinary records. Overhead shelves provide a place for hats and helmets, a few essential books, and other supplies.

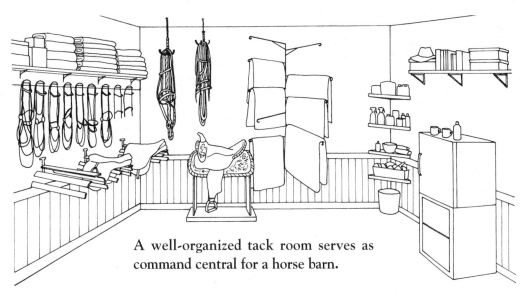

A well-organized tack room serves as command central for a horse barn.

Saddle Storage

Saddles can be stored on floor stands or on wall-mounted racks. Saddle racks should allow as much air as possible to dry the underside of the saddle. Wall-mounted racks free up valuable floor space that can be used to store such items as a vacuum and grooming totes. Saddles on wall racks are also less vulnerable to damage from rodents. A free-standing rack, on the other hand, can be easily moved if necessary, and can serve as a saddle cleaning and repair stand.

3.2 WALL-MOUNTED SADDLE RACKS
A wall-mounted saddle rack can be safely secured anywhere on this ¾-inch (19.1-mm) pine paneling. For thinner wall material, fasteners should go through the wall into a stud.

3.1 SADDLE STANDS AND RACKS
These breathable nylon saddle covers protect leather from dust and dirt while allowing ventilation. You can see here how much floor space could be reclaimed if another row of wall-mounted saddle racks were to replace the bulky wooden saddle stands below. Also, note how the numerous slats on the stands compress the sheepskin lining and prevent it from drying.

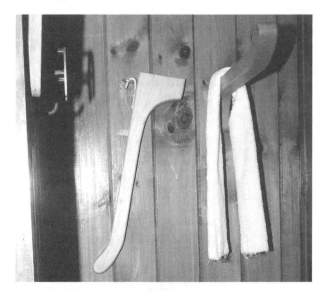

3.3 COLLAPSIBLE SADDLE RACK
Here's a simple English saddle rack consisting of a single piece of wood that attaches by a hook to a screw eye in the wall. When not in use, it hangs flat against the wall or can be removed altogether.

Bridle and Blanket Storage

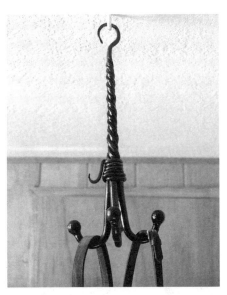

3.4 BRIDLE RACK

Having a separate hanger for each bridle makes it easier to find the bridle you're looking for, and you can retrieve an individual bridle without pawing through a tangled mess. Bridle holders should be rustproof and contoured to hold the shape of the headstall as it dries. Avoid hanging bridles on a single nail, for example, which would make a crease in the leather that could cause it to crack.

3.5 TACK HOOK

This hand-forged tack hook adds a touch of elegance as well as utility. Make sure when hanging tack hooks from the ceiling that they are secured into a solid framing member and not merely into the ceiling cover material.

3.6 BLANKET RODS

This swiveling blanket-rod system is an ideal way to organize a group of blankets. (See chapter 5.)

BLANKET STORAGE TIP

Roll clean winter blankets and hoods, and store them in an out-of-the-way, clean, rodent-free area above cabinets to make the most of available space.

Plumbing and Refrigeration

Plumbing should be protected from freezing yet also needs to be accessible for repairs. Some supplies must be protected from freezing while others need to be shielded from heat and sunlight.

3.7 PLUMBING

This cabinet below a sink holds a small water heater and pressure tank. Heat escaping from the water heater keeps the contents of the cabinet from freezing without an additional heat source. Other ways of providing hot water without installing a huge water heater include a small flash water heater, an on-demand heater, and a timer-actuated heater. In addition, an existing hot water heater can be connected to an on-off switch to heat water only when you need it.

SHELVING TIPS

▶ Cupboards are useful for organizing everything from shampoos to towels to liniment.

▶ Add a shelf above your washer/dryer to make a perfect place for storing detergents and other laundry products. (See photo 3.14.)

▶ Add a shelf close to the ceiling around the perimeter of your tack room to store seasonal blankets in an out-of-the-way place.

3.8 VET SUPPLY REFRIGERATOR

A small refrigerator like this one is relatively inexpensive, doesn't take up much room, and is ideal for storing penicillin, vaccines, and beverages.

Ventilation and Heat

3.9 TEMPERATURE CONTROL

Heaters can be used to warm the tack room so water pipes don't burst and medications don't freeze, and they can make the room a comfortable place to work on blustery winter days. The milk house heater on the left heats up quickly and has a fan that directs the warm air. The oil-filled radiator heater on the right takes longer to heat up but provides a more even and energy-efficient heat. A fan is nice on hot summer days to move air through the tack room or for keeping flies off a horse in the grooming area.

3.10 FLOOR VENT

A tack room, especially one with no windows, can get stale and stuffy. An operable wall vent located close to the floor opposite a screened door or window lets fresh air circulate naturally. Install a small screen behind the vent to keep insects out.

3.11 CEILING-VENT FAN

This ceiling-vent fan is used in the summer to draw hot air out of the tack room, which then draws fresh air into the room through the floor vent (see photo 3.10). The fan is controlled by a timer set to come on at 11 P.M. and turn off at 5 A.M., thereby filling the tack room with cool night air. The tack room is well insulated, and will stay cool until late afternoon.

Repair and Cleaning Areas

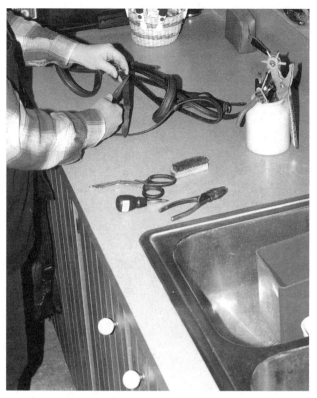

3.12 TACK REPAIR AREA

A well-lighted area such as a small table or counter is useful for making small tack repairs and adjustments. If the tack room is shared by many people, try to provide a cleaning and repair area away from main traffic patterns, perhaps in a separate room.

3.13 SADDLE-CLEANING STATION

This ingenious wooden stand holds an English saddle at just the right height for easy cleaning of the outer surfaces. The V-shaped front of the stand holds the saddle securely upside down for cleaning the bottom panels. Containers beneath the stand hold cleaning supplies.

3.14 LAUNDRY CENTER

Few people want to wash horse laundry in their home machines, and trips to the laundromat can take a chunk out of your riding schedule. This washer and dryer were moved to the tack room after serving their tour of duty in the residence, and are invaluable for cleaning boots, wraps, cloths, and sponges. They can handle most coolers and sheets and even some lightweight winter blankets (see chapter 16 for blanket washing tips).

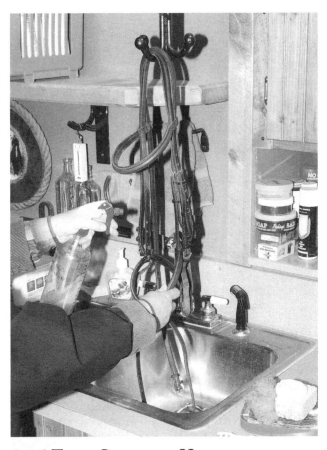

3.15 SCRUB SINK

Handling horses and tack can be dirty work. A sink with hot and cold water allows you to wash your hands as often as necessary, which can help prevent spreading skin infections among horses.

3.16 TACK CLEANING HOOK

This tack hook positioned over a sink makes it easy to clean a bridle after each use and is a great place for taking bridles apart for a thorough cleaning when time allows. It's mounted high enough so that when it's empty, you won't hit your head on it when using the sink.

✦ WORK AREAS ✦

A work space should have some means of confining your horse but still provide sufficient room to accomplish your tasks. Tools should be within easy reach, yet stored safely out of the way so they aren't run into or knocked over by either you or your horse. The flooring of a particular work space depends on the purpose of the work area, whether the area is protected from moisture, and whether it is indoors or outdoors. Level dirt is easy, but a pawing horse learning to stand tied can quickly make a crater that will turn into a pond during wet weather. Concrete is durable, but smooth concrete can be very slippery, and rough concrete can cause excessive wear to hooves and horseshoes. Rubber mats over concrete provide cushion, stability, and durability whether inside or out, and are an ideal choice for almost any work area (see also chapter 1).

A Safe Place to Tie

Never underestimate the power of a horse. Wherever you tie a horse, whether to a post, a rail, a ring, or something else, it should be secure and strong enough not to break or come apart if the horse pulls back. Tying to a board fence, for example, has been the cause of many serious injuries to both horses and handlers, when the board either broke or was pulled loose from the posts. If the loose board remains tied to the lead rope, it will "chase" the terrified horse, leading to both physical and psychological injury.

The higher a horse is tied, the less leverage he has to pull back, so the point where the rope connects should be at least as high as the horse's withers. Wooden posts and rails can quickly become marred and eventually weakened by chewing horses. A tie post should be in the ground at least 4 feet (1.2 m) deep and anchored securely, preferably in concrete, so it doesn't work loose. Wooden tie posts should be at least 6 inches (15.2 cm) in diameter for a 1,000-pound (454-kg) horse. If snugged tightly on a wooden post, a rope will usually stay in place above the horse's withers. On the smoother surface of a steel post, a loop can be welded to the post to keep the rope from slipping down.

When attaching a tie ring to a wall or post, use lag screws (bolts with a wood-screw thread and hexagonal head) at least 5/16 inch (7.9 mm) in diameter and long enough to go through the wall surface and 2½ inches (6.4 cm) into the wall framing. On a wall, the tie ring must be located on a stud or other secure framing member within the wall.

There should be at least one tie ring in every stall, so the horse can be tied when cleaning the stall or kept from moving about when grooming or doctoring.

Cross ties are two ropes on either side of the horse that attach to the side rings of the halter. Horses have quite a bit of freedom to move forward and backward in cross ties; they must be trained to stand quietly. Very wide cross ties should be mounted high to keep the ropes out of the way, while narrow cross ties can be mounted level with a horse's poll.

Tying

4.1 HITCH RAIL
A hitch rail is the traditional place to tie a horse. This Southwest model is plenty stout but too low to be able to tie at the height of the withers. Also, the wood will likely need frequent treatment with an antichew product (see chapter 8) to keep horses from gnawing on it.

4.2 TRI-LEVEL HITCH RAIL
This tri-level hitch rail has loops welded to the steel pipe to keep the rope from sliding back and forth on the pipe. The two lower rails keep a horse from going underneath and provide a place to hang blankets to dry. Chewing horses can scrape the paint off, but won't weaken the steel pipe.

4.3 TYING PROBLEMS
Although this horse has been wisely tied to the post, about withers height, rather than to the board fence, there are several potential problems here. First, the horse could simply lift the rope up off the post and leave. Second, judging by the amount of chewing that has gone on, this is lucious, untreated wood. Third, the ceramic electric fence insulator on the top of the post is a hazard, and since it is no longer in use, it should be removed.

Tying (continued)

4.4 CROSS TIES

The distance between the cross tie rings should be no wider than 11 feet (3.4 m) so a horse cannot turn around (see photo 4.8). If a horse gets turned around in cross ties the ropes will cross his face, twisting his halter, and will likely cause the horse to panic and pull back, which can cause very serious injury to his face and possibly his body if the cross ties break or he falls. Never leave a horse unattended in cross ties.

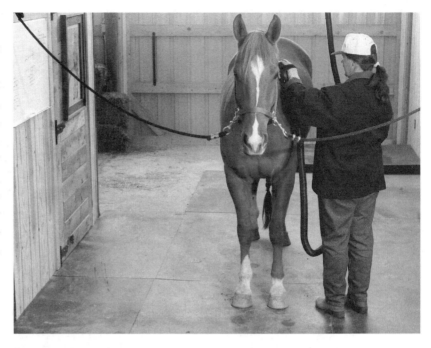

4.5 TYING SAFETY

A panic snap *(left)* is a snap that can be easily opened by sliding a metal collar at the base of the snap. This type of snap can even be released while there is tension on the rope. Using panic snaps on the ends of cross tie ropes will allow you to quickly release a horse if he gets in a bind. Ropes should always be tied with a quick-release knot *(center)*. You should always have a sharp pocket knife *(right)* handy in case the snap or knot doesn't release.

4.6 CROSS TIE RING

This sturdy cross tie ring is connected to the 2 x 6 stud within the wall with ⅜-inch by 3-inch (9.7-mm by 7.6-cm) lag screws.

Grooming Area

The grooming area should be uncluttered so a person can move all around the horse without tripping and stepping over things.

4.7 BOARDING STABLE

This open grooming area at a large barn has an easy-to-clean concrete floor and several tie posts. The tack closets along the walls are convenient; the space above them is used for storing buckets and other large items. A tie post in the open like this allows too much freedom of movement for some horses and could lead to problems.

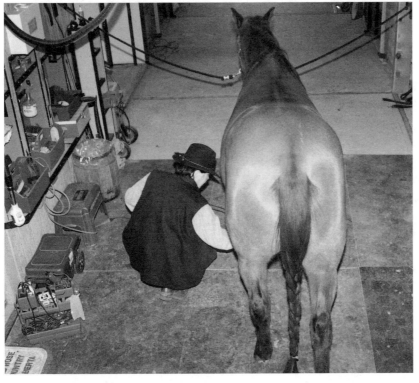

4.8 PRIVATE BARN

Smooth rubber mats are quiet and easy to clean, and provide cushion for both horse and groom. Light coming from the sides, rather than from directly overhead, reduces shadows and helps the groom see all parts of the horse more clearly. The width of this grooming space is 11 feet (3.4 m), which is the maximum safe span for cross ties. (See photos 4.9–4.11 for details.)

Grooming Area (continued)

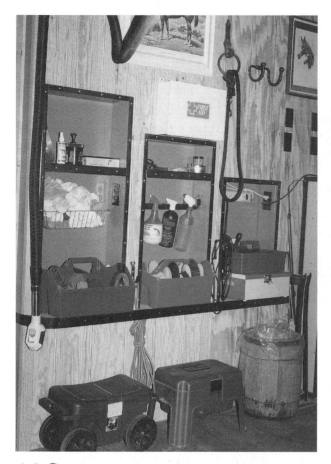

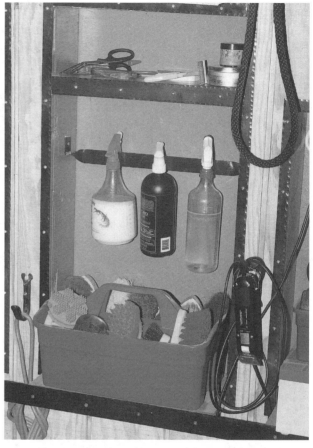

4.9 CUBBYHOLES

The wall of this grooming area is framed of 2 x 6 studs, between which shelves are attached. These "cubbyholes" provide a space where grooming tools and supplies are accessible yet out of the way. Electrical outlets are safely located inside the recessed areas. Hanging down on the left is the hose from a central vacuum. Above the middle cubbyhole is a first-aid kit and above that, a cross tie ring and bridle hooks.

    The wider bottom shelves hold portable totes. A tote can be used for only brushes, for example, or each horse can have a separate grooming tote with his name on it. The latter method lessens the chance of spreading skin infections between horses. On the floor are a rolling stool, for sitting on when clipping a horse's lower legs, and a combination storage/step stool for working on manes of tall horses. The wooden keg holds a plastic trash bag.

4.10 GROOMING SUPPLIES

This detail of the center cubbyhole shows the top shelf being used for scissors, hoof picks, a small jar of peanut butter for mousetrap bait, and other small items. Spray bottles hang on a metal strap across the center. A tote holds grooming tools, and clippers hang ready for instant touch-ups.

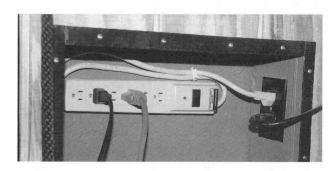

4.11 SURGE STRIP

A surge strip attached to the wall provides additional outlets for clippers and vacuums.

Farrier Work Space

The space where your farrier works should be protected from rain, wind, and direct sun. A farrier works out of his truck, so the shoeing area should be located where a vehicle can be backed close to it. The floor should be level, smooth, and uncluttered.

It should have good traction and be kept free of gravel and other debris. It's best if the area is lighted from the sides or corners, rather than from overhead. Most farriers use electric tools, so access to a 110-volt outlet is essential.

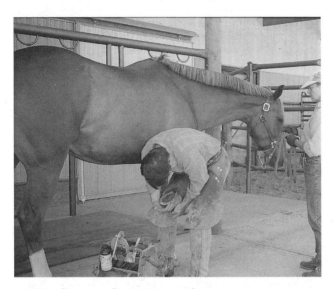

4.12 GOOD OUTDOOR SITE

This clean concrete slab is under a roof overhang that protects it from rain and snow and provides shade in the afternoon. This farrier has asked the person to hold the horse alongside the hitch rail to limit the horse's movement. Some farriers would choose to tie the horse to the hitch rail.

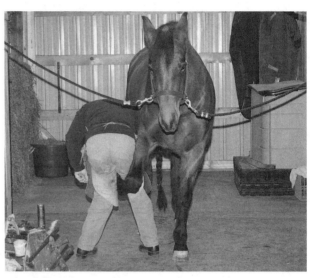

4.13 GOOD INDOOR SITE

Horseshoer's heaven: a well-lighted, quiet area with rubber mats to work on and a horse that stands centered and calm. The farrier's truck is parked just outside the large sliding door behind the horse. On hot summer days the door can be left open to let in a breeze.

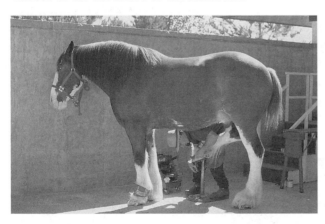

4.14 UNCOVERED AREA

This shoeing area has a smooth concrete floor and a sturdy tie ring, but because it is uncovered, it is sometimes too hot at midday.

4.15 BAD SHOEING AREA

This is a *good* example of a *bad* place to shoe. There is no protection from sun and rain, the area is cluttered, the tie rail is weak and low, and the ground is covered with duck droppings.

Vet Area

A veterinarian, like a farrier, should be able to get his or her truck close to the work space in the barn. Many veterinary procedures involve washing various parts of the horse, so access to hot and cold water is a big plus. A counter or table, close to where the vet is working, is very useful for setting tools and keeping supplies at hand without having to set them on the floor. The footing should be solid, with good traction, and sloped to drain water away from the work area so the vet doesn't have to stand in water. Good lighting is especially important for vet work. Besides ample permanent lights, portable lights should be available to spotlight certain areas on the horse. Plenty of outlets installed in convenient locations will make it easier for the vet to use clippers, ultrasound, and other equipment. The safer and more convenient you make the work area, the better job your vet will be able to do.

4.16 PROFESSIONAL STOCKS
A professional palpation chute (stocks) is designed to contain a horse safely while allowing a veterinarian to work on it. The solid doors on the front and rear of the stocks shown here protect a person from being kicked or struck. The floor slopes to a drain located between the two stocks. With doors on both ends, the horse can be led in and then out of these stocks without having to back up.

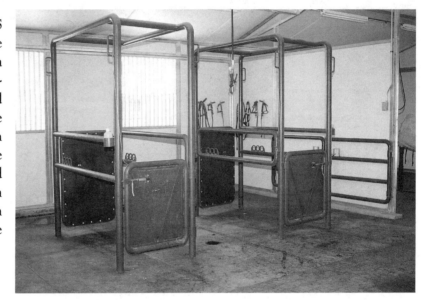

4.17 OUTDOOR STOCKS
These simple outdoor stocks are mounted on a concrete slab in the shade of a tree. The open design makes them useful not only for vet work, but for bathing a horse as well. The horse must be backed out of these stocks.

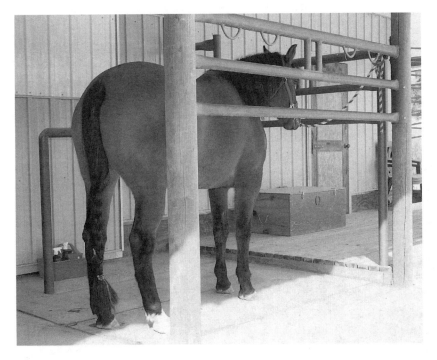

4.18 LOOSE STOCKS (SIDE VIEW)

"Loose stocks" are roomier and multipurpose. These loose stocks are located next to a barn under a roof overhang, and are designed as a place to train young horses to stand quietly and to let a horse dry after a workout or a bath. They can also be used for vet work.

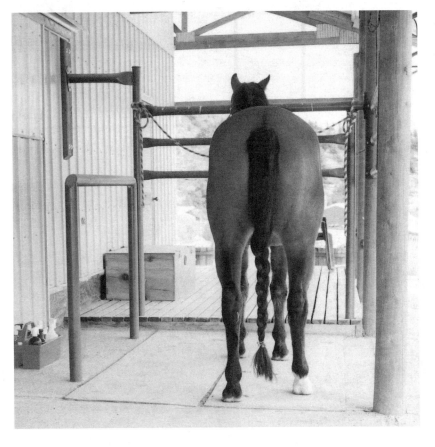

4.19 LOOSE STOCKS (REAR VIEW)

From the rear, you can see how the cross ties and pipe rails allow some movement, but keep the horse from swinging too far sideways. There's enough room in the loose stocks for a farrier, vet, or groom to work on a horse. A thin rubber mat over the textured concrete prevents excess hoof wear from pawing.

Wash Rack

A wash rack requires hot and cold water and solid footing with good traction. The floor should be sloped to drain so the handler isn't standing in water. The walls should be waterproof so the wall framing doesn't rot. Light fixtures should be placed so light comes from the sides, not centered overhead, and the fixtures should be sealed, as should outlets and switches. The area should be protected from breezes so a wet horse doesn't get chilled. Because horses sometimes move quickly and knock into things when startled by water spray, there should be no protrusions such as water faucets that could injure or be damaged by a horse. When there's no room inside the barn, a wash rack can be located outdoors. An outdoor wash area is most useful in consistently warm climates.

4.20 OUTDOOR WASH RACK

This outdoor wash rack is protected from the wind and has rubber mats over concrete. Instead of installing a drain, a concrete trough was formed off the front corner of the floor. (See photo 4.17 for another outdoor wash rack option.)

4.21 WASH RACK/ VET COMBO

This wash rack doubles as a vet area. Hot and cold water are available from a sink as well as from a hose connected to freeze-proof hydrants. There are cross ties at both ends of the wash rack so the vet can position the portion of the horse she is working on near the sink and counter. Here a mare is positioned for the vet to palpate.

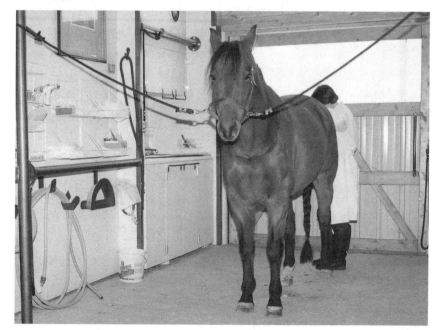

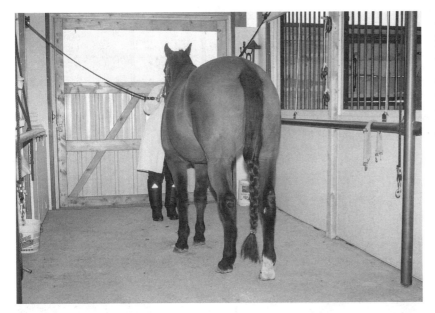

4.22 TWO-WAY CROSS TIES

Now the horse is facing the other way and is tied to the other set of cross ties to have her teeth floated. Pipe side rails keep a horse from crashing into the walls. The sliding door at the back of the wash rack allows easy access to the vet's truck.

4.23 WASH RACK/VET SINK

The one-piece stainless steel sink and counter provides the veterinarian a place for washing tools and for setting equipment while working on a horse. Upper and lower cabinets store bandaging materials, towels, soap, and other supplies. The cupboard handles double as towel bars.

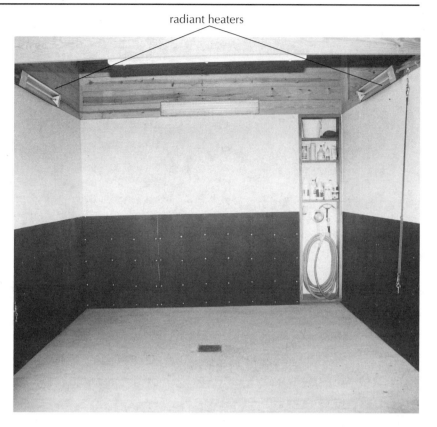

radiant heaters

4.24 PADDED WASH RACK

This spacious wash rack has a textured concrete floor that slopes to a drain in the center. White fiberglass-reinforced plastic (FRP) on the upper half of the walls helps illuminate the stall by reflecting light. Rubber mats, ¾ inch (19.1 mm) thick, on the lower portion of the walls are durable enough to be unaffected even by direct kicks. Cross ties connected high on the walls are positioned so the horse faces the barn aisle. More tie rings on the walls would allow greater flexibility when positioning horses. The faucet, a single-knob shower-type, along with a hose hanger and shelves for shampoos and tools, are all recessed into the back wall at the corner, well out of the way of any horse contact. There is a radiant heater mounted high on each side. Fluorescent lights with vapor covers are located on the front beam and the back wall.

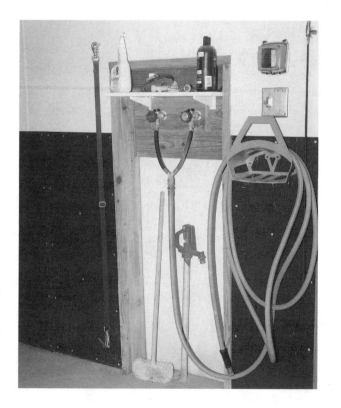

4.25 GOOD AND BAD

In this wash rack, the hydrant and faucets are recessed but the hose hanger and the shelf are not, so they could be bumped by a horse. Note the waterproof covers on both the light switch and heater control. The hot and cold faucets are joined by a Y connector.

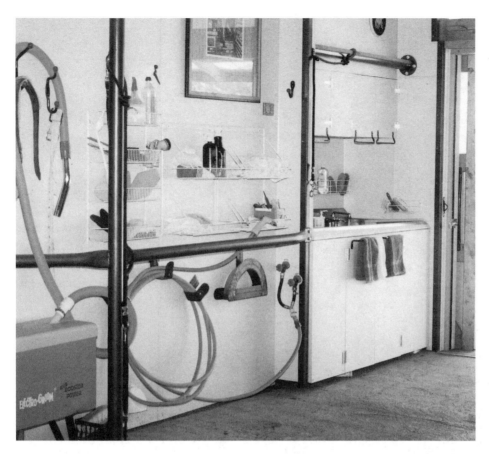

4.26 BUMPER BAR

In this wash rack, a heavy horizontal steel pipe prevents a horse from swinging over and contacting the wire racks on the wall and the hose hanger and faucet handles below.

4.27 FLOOR TEXTURE AND DRAIN

When this concrete floor was poured, before it set, a stiff broom was dragged across the surface to make a textured, wavy pattern for good traction when wet. The rough concrete surface grabs most of the hair, so it can be swept up rather than flushed down the drain. A 5-inch (12.7-cm) diameter plastic drain cover also keeps hair and debris from clogging the drain. Larger drain covers are more likely to be broken by the weight of a horse.

TIP

Waterproof outlet covers are a must for wash racks. In order to be waterproof, the covers must be closed!

Wash Rack (continued)

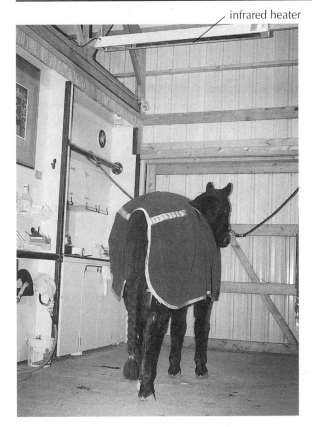

infrared heater

4.28 HEATED WASH RACK

Along with the aid of a special sheet called a cooler, an infrared heater is useful for allowing a horse to gradually cool down after a vigorous workout in very cold weather. This 4-foot (1.2-m) long electric infrared heater (see Appendix) mounted over the wash area makes bathing horses in colder temperatures safer, without risk of chilling. Infrared or radiant energy produces heating effects similar to the sun but without the visible light and UV rays. It heats objects, not the air. A controller for the heater is mounted in the cabinet beneath the sink.

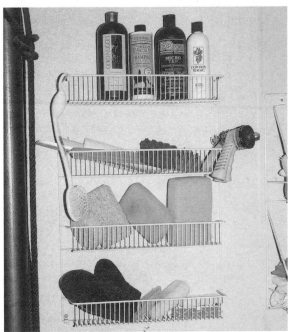

4.29 BATHING SUPPLIES
Rubber-coated wire racks mounted to the wall keep shampoos, sponges, and other wash rack items accessible, yet out of the way.

4.30 TACK CLEANING AREA
A tack hook suspended from the ceiling of this wash rack provides an excellent place to spray off large or dirty items such as surcingles and cinches.

5
✦ STORAGE AREAS ✦

"A place for everything and everything in its place." Apply this maxim to your barn, and it will be safer for you and your horses and a more pleasant and efficient place to work. Tools, equipment, and other items scattered about the barn can be stepped on, tripped over, or snagged as you lead your horse or hustle to do chores. If your horse knocks over a rake, for example, he might step on it and puncture the sole of his hoof. If he gets caught on an extension cord or piece of hanging tack, it could scare him and he might bolt forward or jump sideways, injuring himself, you, or other persons nearby. Barn aisles should be kept clear at all times so horses can be led safely and directly without having to negotiate an obstacle course.

Store tools and supplies so they are secure and out of the way, yet accessible. Park large items like carts, barrels, and trunks away from the main travel lanes in a separate alcove or an extra stall. Each piece of equipment, no matter how big or small, should have its own storage place. Require that items be returned to their proper place after use. Such organization eliminates time wasted searching for tools and supplies. Keeping close track of your tools and equipment reduces the likelihood that they will be misplaced or lost, thus minimizing unnecessary replacement costs. A mechanic's toolbox or drawers with interior dividers, located under a workbench, keeps small tools, such as pliers, screwdrivers, and wrenches, organized and clean.

Store grooming tools that are used often on shelves or in a cabinet close to the grooming area. Another option is to organize tools in totes that can be carried to the grooming area as needed.

Blankets, ropes, and other items containing natural fibers can be damaged by moisture, dirt, moths, rodents, and other animals. Keep such items clean and store them in tight closets or trunks. Check them periodically to make sure they aren't being gnawed or used for nest materials by uninvited guests. Mothballs in the storage area can discourage insects and rodents.

Rods or bars mounted close to the grooming and tacking area make a convenient place to store saddle blankets between uses. Store large, frequently used horse blankets on a bar mounted near each horse's stall. That way the blankets are ready for immediate use and you know which blanket goes with which horse.

Hay and bedding are very flammable; it's best to store large quantities in a building separate from the barn. Small amounts, enough to last a week or two, can be moved into the barn as needed. Hay and bedding should be stored out of the way in an extra stall or an alcove. Stack hay and bedding on wooden pallets, if necessary, to keep ground moisture from causing spoilage. Loose hay can be slippery underfoot, so sweep the floor regularly to prevent mishaps.

Tool Storage

A place especially for tools, such as an extra stall, small room, or nook, helps keep the main barn aisle uncluttered. The tool room should be away from traffic areas and out of reach of horses, to prevent injury to horses and damage to tools.

5.1 NOT SO GOOD
Leaning long tools such as forks, rakes, shovels, and brooms against a wall takes up precious floor space, and can make it awkward or dangerous to locate specific tools when you need them.

5.2 VERY GOOD
Instead, use wall hooks and brackets, which provide a convenient way to store long tools so they're out of the way, yet visible and easily accessible.

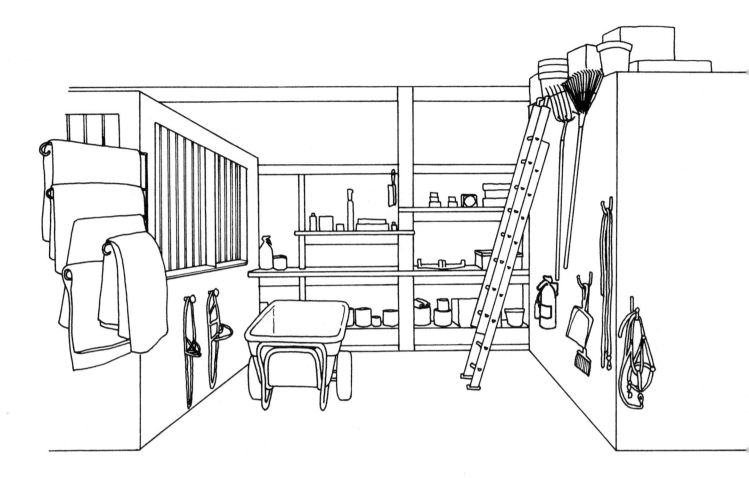

A tool room should be located away from horse traffic. This three-sided tool room is 8 feet (2.4 m) wide by 10 feet (3 m) deep. It is bordered by a stall on the left and a feed room on the right, and opens onto the main aisle of the barn. It's an appropriate size for a barn that houses from two to eight horses. There is room to park a manure cart out of the way, with plenty of room left over for storage and work space.

Swiveling blanket rods are mounted on the corner, within a few steps of the grooming area and the stall. They're great for drying saddle blankets between uses and for storing extra fly sheets and stable sheets.

Hooks on the wall below the stall grille can be used to store extra halters, ropes, or extension cords. Here, proper spacing of the grille bars is particulary important, as the horse should not be able to reach through with his muzzle and grab hanging items or items on the tool bench.

The bench along the back wall serves as a mini-shop for small repair jobs and other barn-maintenance tasks. Shelves above the bench provide handy storage for screws and nails, hand tools, spray lubricants, and other small items. Larger items such as cans of paint, antichew products, portable lights, and stall freshener products can be stored under the bench.

The ladder on the right (stored elsewhere) is positioned to access the loft over the feed room (see page 42). Because the exterior wall of the feed room is ⅝-inch (1.6-cm) thick plywood, hooks can be mounted anywhere on its surface for light items such as forks, rakes, ropes, cords, and halters. (Be sure the mounting screws aren't so long that the ends project into the feed room where they could cut someone.)

Forks and rakes are hung with their business ends up so they take up the least amount of room and are easy to grab by the handles when needed. A large dust pan and hand broom hang near the aisle, handy for needed clean-ups. Extra halters and lead ropes hang near the aisle for instant access.

Grooming Supply Storage

5.3 BARN AISLE STORAGE

These shelves, 11 inches (27.9 cm) high and 10 inches (25.4 cm) deep, provide a convenient place to store fly sprays and grooming products in the barn aisle, where they're most frequently needed. Sunlight can reduce the effectiveness of some products, however, and in cooler climates, products affected by freezing would need to be moved to a temperature-controlled area in the winter. A storage bin with a flat lid, lower right, invariably becomes a catchall, which makes opening the bin inconvenient. (See also photos 4.9 and 4.10.)

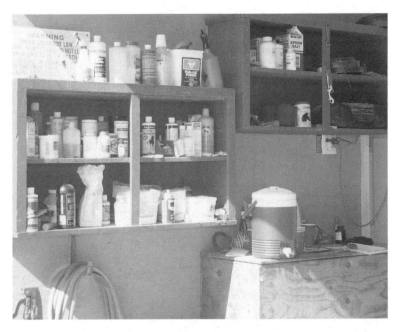

5.4 SMALL-ITEM ORGANIZER

This rolling plastic bin of transparent drawers is one way to keep track of small items such as buckles, snaps, mane bands, and small tools, like pliers, screwdrivers, and tape measures.

STORAGE TIP

A dresser or chest of drawers is a good place to store clippers and blades, bandaging materials, and leg and tail wraps.

Hay, Bedding, and Blanket Storage

5.5 HAY AND BEDDING
This three-stall shed is 24 feet (7.3 m) wide by 16 feet (4.9 m) deep, and is used for storing hay, straw, and sawdust. The open front faces southeast, away from prevailing winds.

5.6 STALL BLANKET ROD
Storing a blanket and halter on each stall door ensures that they are at hand when you need them. This is also a good way to designate which blanket and halter go with which horse.

5.7 SWIVELING BLANKET RODS
This set of blanket rods is located near the grooming area, just three steps from the horse. The rods provide a handy place to store saddle blankets, sheets, and coolers where they can dry and air out between uses. The rods swivel so they can be staggered as needed or completely moved out of the aisle and the traffic area.

Coils

Ropes and extension cords should be neatly coiled and hung on their designated hooks after each use. You'll know where to find them when needed and you and your horses won't be tripping over them.

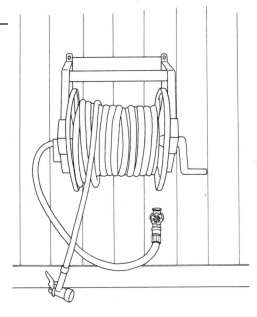

Hoses can be neatly coiled and hung on a hanger (see photo 13.16), but it's faster and more convenient to roll a hose onto a special spool like this one.

Lofts

A loft can be a good place to store items that are seldom used. A loft can be a formal second story, such as is often used for storing hay, or it can be the floor space created above ceilings, of a feed or tack room, for instance. Some ceilings aren't built to carry additional weight, so make sure the framing is strong enough to support you and the items you are storing. Lofts are notorious for gathering dust, so you might want to put stored items in boxes or plastic bags to keep them clean. It's a good idea to do "loft cleaning" at least once a year to minimize fire hazards, weed out unwanted items, reorganize the stuff you want to keep, and check for rodent and bird infestation.

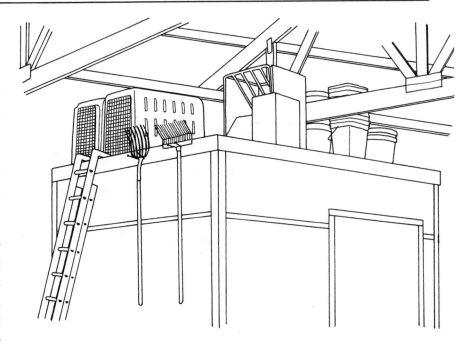

Even though a loft may not have much headroom when you are up there sorting, it can still provide storage for seldom used items such as dog kennels, extra feeders, an emergency propane heater, and empty buckets.

✦ TOOLS ✦

Ideally, a barn should be equipped with a set of essential tools that stay in the barn, not scattered between truck, garage, and shop. That way, you'll be sure to have them when you need them. Know which tools you need to outfit your barn and buy the best you can afford. Quality tools are a good investment, and the better you care for your tools by keeping them clean, dry, and in good repair, the longer they will last.

Hand Tools

You can waste a lot of time hunting for small tools if they get misplaced. Mark the handles of small tools with bright colors to make them more visible and easier to find. Options include paint, tape, and a colored rubber coating (available at hardware stores) that also improves grip.

A tamper is used for compacting dirt and gravel when filling in holes in pens and stalls, or when preparing a floor for rubber mats. Tampers are available in hardware, building supply, and farm stores or can be made from an old steel railroad tie plate and a section of pipe.

It's good to have a crowbar and a smaller flat bar around when you need to pull a nail, pry two panels apart, lift up the corner of a stall mat, or break a hole in the ice on a water tub. A 4-pound (1.8-kg) sledgehammer packs a wallop when you need to drive steel stakes into the ground to secure a railroad tie or pen in place (see chapter 7). A 16-ounce (454-g) carpenter's claw hammer is a good all-around hammer for driving nails, closing paint cans, and sundry other common tasks. A very small ball-peen hammer is useful for repairing tack items.

TOOL TIP

Especially in dry climates, wooden handles dry out and shrink and hammer heads become loose. To keep a hammer handle tight, stand the hammer on its head in a shallow container and pour in enough antifreeze to cover the head. Soak for two days.

Warning: Antifreeze has a sweet taste that animals find appealing and it is very toxic, so keep it out of the reach of all animals and children.

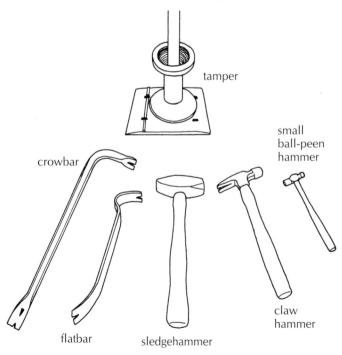

tamper

crowbar

small ball-peen hammer

flatbar

sledgehammer

claw hammer

A rotating leather punch is the best tool for punching additional holes in leather tack.

Locking pliers are great because they will clamp and hold onto an item, such as a nut or bolt, without you having to exert constant pressure. Regular pliers are good for such jobs as opening cans with small screw-on caps and twisting light wire to fasten things together. Needle-nose or long-nose pliers are good when regular pliers are too wide, such as for some tack repair jobs.

When you need to cut wire, side cutter or end cutter pliers are the best tool for the job. A fencing tool is very handy to have around the barn because it can cut wire, firmly grab objects at several different places, and drive and remove nails or staples.

A small awl can be used for starting screw holes in wood. Heat the tip with a candle or small torch to melt holes in nylon straps.

A straight screwdriver is needed for slot-head screws. A small magnetic screwdriver is useful for small jobs, such as working on clippers, and for retrieving dropped screws from tight places. A Phillips screwdriver is essential for star-shaped Phillips head screws. Combination screwdrivers have interchangeable bits.

A ⅜-inch (9.7-mm) cordless drill with Phillips, straight, and hex bits (hex-head screws are used for steel roofing and siding) is indispensable for quickly driving and removing screws — anytime, anywhere!

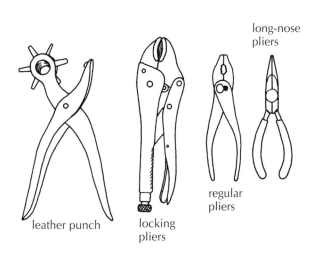

long-nose pliers

leather punch

locking pliers

regular pliers

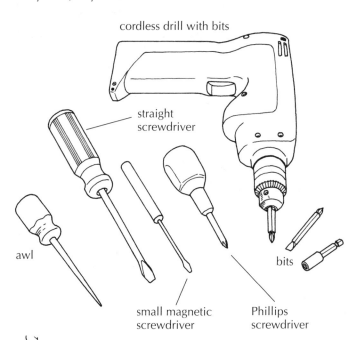

cordless drill with bits

straight screwdriver

awl

small magnetic screwdriver

Phillips screwdriver

bits

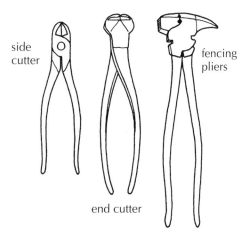

side cutter

fencing pliers

end cutter

This strong magnet on a cord is for picking up horseshoe nails off the floor or for finding small items such as screws or nails in a horse's bedding. Similar magnets attached to a long handle can be found in some building supply and hardware stores. (See Appendix.)

An emergency farrier kit is invaluable when you need to pull a loose shoe. Trying to pry off a loose shoe with a screwdriver can injure the horse and damage the hoof.

A clinch cutter is used like a chisel to cut or pry open the nail clinches. A hammer such as a carpenter's hammer is used to tap the clinch cutter. Pull-offs are used to pull loose nails and pry off the shoe (pull-offs resemble hoof nippers but are wider and not as sharp). A crease nail puller is used to pull individual nails by grabbing the heads.

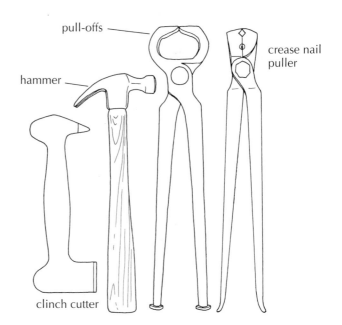

Forks and Rakes

Plastic stall forks are surprisingly durable, lighter than steel forks, and best suited for cleaning manure off of dirt or small gravel. They are not well suited for frozen or irregular ground. A basket fork with a high back and sides holds more manure.

A steel silage fork is best for cleaning irregular or frozen ground because the smooth steel tines slide easily over the ground. It's better than plastic for sifting bedding because of the smoother, slightly wider-spaced tines. Don't confuse it with light-duty metal "stall forks" with springy wire tines that catch on everything and bend.

An old-fashioned hay fork is handy for loading composted manure into a spreader and for cleaning stalls bedded with straw and the hay barn floor.

A plastic or bamboo leaf rake is the best tool for gathering up manure that's been scattered about the pen by the horses.

A landscape rake is invaluable for regular leveling of pens and runs, filling in small holes and keeping the footing evenly distributed. Commercial models often are smooth on one edge and have wide, heavy tines on the other edge.

Brooms and Shovels

A push broom with stiff plastic bristles is used to sweep gravel off concrete barn approaches and off rubber mats in the horses' outdoor feeding areas.

A wider, softer-bristled push broom makes quick work of sweeping the textured concrete and rubber mats in the barn aisle.

An angle-cut broom with soft bristles is primarily for the wooden floor in the tack room, but is also used for the smooth concrete floor in the feed room and for preliminary sweeping of nooks and crannies in the barn aisle, before sweeping with a big broom.

The short broom is a recycled long broom now used as a hand broom in the grooming area for sweeping manure onto the dustpan. Adding a cord through the handle allows the hand broom to hang on a hook with the dustpan, within easy reach of the output end of the horse. A wide, lightweight aluminum dustpan is much easier to wield than a shovel, and the bending over required to use it can be used as a stretching exercise.

A 2-inch (5.1-cm) paintbrush is used for dusting narrow places on shelves and tools and for cleaning the filter on the horse vacuum.

A bench broom is used to sweep off the tops of trunks, clean out corners in the tack room and feed room, and dust off the plastic lawn chairs on the porch of the barn (which tells you how often the chairs are used).

In snow country, a snow shovel could be needed just to get to the barn, so keep one at the house, too. An aluminum scoop shovel is easy to handle because of its light weight. It's good for deep snow, scooping manure off rubber mats and concrete, and handling bulk bedding, such as sawdust and shavings.

A steel scoop shovel is quite a bit heavier than the aluminum one, but also sturdier. It is good for moving a lot of sand or gravel if you're a strong person.

The flat shovel is a good choice for shoveling gravel and for leveling pens. It won't wear you out as quickly as the scoop shovel will.

A common pointed shovel is the one to grab when you need to dig a hole or trench or move a pile of sand or dirt.

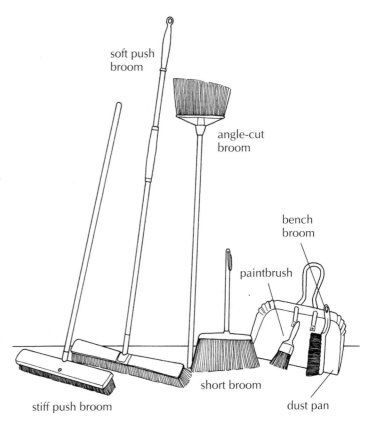

soft push broom

angle-cut broom

bench broom

paintbrush

stiff push broom

short broom

dust pan

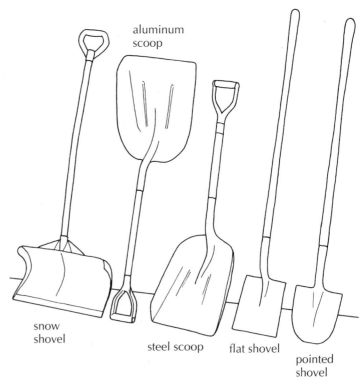

aluminum scoop

snow shovel

steel scoop

flat shovel

pointed shovel

Carts

The small plastic garden cart (Rubbermaid Specialty Products, Wooster, OH) is very lightweight and surprisingly tough. It holds one bale of hay stacked in two sections. Its hard plastic tires are very noisy on concrete, however. (The other carts featured in photo 6.2 have rubber tires.)

The plastic cart on a lightweight tubular frame (Smart Cart, True Engineering, Windham, ME) is very well balanced. The easiest cart to maneuver and dump, it makes an excellent manure cart.

The bicycle tires on the plywood cart (Homestead Carts, Dallas, OR) make it easy to roll. It is too wide for some doorways and not easy to dump and so is more suited for feeding hay than gathering and dumping manure. It holds two bales of hay. The steel lining of the bed cleans easily.

The tractor cart (Rubbermaid) is pulled with a small vehicle like a garden tractor. It is heavy-duty, and is especially useful for hauling heavy loads like rocks, sand, and dirt.

The large plastic cart (Rubbermaid) with rubber tires easily holds one bale of hay or a generous load of manure. It's easy to push and maneuver, but the lack of handholds underneath the cart make it difficult to tip forward to dump.

6.1 The hardest part of doing chores can be dumping a heavy cartload of manure. The design of this lightweight cart is better than most, because once you tip the cart forward, you can grab the tubular legs and pull the cart up and backward to finish dumping.

6.2 From left to right: small plastic cart, plastic cart on tubular frame, plywood garden cart, tractor cart, large plastic cart.

Equipment

A standard 22-inch (55.9-cm) lawn mower can be used on level areas close to the barn and pens, where the grass is kept trimmed fairly short.

A handheld string trimmer is handy for places a mower can't reach, such as close to buildings and pens. Gas-powered trimmers can go anywhere, but can be noisy. Electric trimmers are quieter, but mobility is restricted by the extension cord.

The gasoline-powered wheeled trimmer is for serious weed cutting (see photo 16.17) and good on irregular ground.

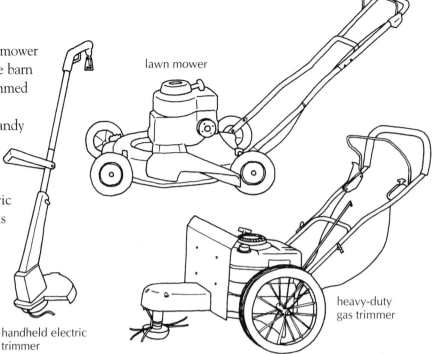

lawn mower

heavy-duty
gas trimmer

handheld electric
trimmer

6.3 HORSE FARM TRACTOR

This 62 horsepower, four-wheel-drive tractor is the ideal size for a small stable. It's compact, maneuverable, and powerful enough for just about any task. The front-end loader can be used for cleaning out pens, loading compost into a spreader, clearing snow, and moving rocks, gravel, dirt, and bedding. Other attachments allow you to mow pastures and weeds, clear brush, work arena footing, and drill postholes.

A small four-wheel utility vehicle or even a refurbished electric golf cart can be very useful for delivering hay and grain, and to haul tools and supplies, as for mending a fence.

✦ TURNOUT AREAS ✦

Turnout areas are pens, runs, corrals, or paddocks where horses are turned out for limited exercise. Pens are usually the size of a large stall, about 16 by 20 feet (4.9 by 6.1 m). A run is longer, say 16 by 60 feet (4.9 by 18.3 m), where a horse can get more exercise by trotting or even loping for a few strides. A corral is a square or round pen of at least 1,600 square feet (148.6 m²). Pens, runs, and corrals are the most common types of turnout areas that are attached to a barn. They are usually devoid of grass due to concentrated horse traffic. A paddock is usually a grassy area, from a half acre to several acres in size, used for short periods of turnout and very limited grazing. Paddocks are used with discretion, so as to preserve the ground cover. A pasture is a larger, well-maintained grassy area that is used mainly for grazing. Especially in hot climates, turnout areas should provide protection from direct sun. A run-in shed or trees are common shade providers.

Above all, turnout areas should be safe. The fencing should be designed for horses, not cattle or pigs. A sturdy 5- to 6-foot (1.5- to 1.8-m) perimeter fence, complete with gates, should encircle the premises to contain horses that escape from stalls or turnout areas (see Security, chapter 14). Areas should be free of junk and litter that could injure a horse by contact or ingestion. Feeders and waterers should be designed for horse safety (see chapters 12 and 13) and should be securely attached or anchored.

Because pens and runs are devoid of grass or other living ground cover, they can easily turn to mud unless covered with a draining footing such as gravel. Repetitive traffic patterns and pawing often create unsightly holes in pens and runs that make cleaning the area difficult. During wet periods, moisture collects in the holes, forming messy puddles. It takes regular effort to keep pens and runs level and dry. Sand is inappropriate footing for a turnout pen if a horse is fed there, because of the danger of colic from ingested sand.

7.1 Horses benefit from regular free exercise in a run or paddock that's large enough to let them kick up their heels. This filly is wearing a breakaway safety halter.

Turnout Pens

7.2 SHORT SOLID WALL
This attractive adobe pen blends with the barn and the Southwest landscape and is safe and maintenance free. However, at just over 4 feet (1.2 m) high, a horse with sufficient jumping ability could easily get over this wall.

7.3 TALL PANELS
Horses are gregarious creatures — they like to hang out together. Even when separated by fences, horses will try to interact by playing, fighting, and mutual grooming. Some socializing is okay, but mutual grooming can destroy manes, tails, and blankets, and aggressive behavior can lead to damaged facilities and injured horses. Ingestion of too much mane or tail hair (such as by a foal) could lead to life-threatening colic. Tall panels, like the 6½-foot-tall (2-m) panels (Priefert Manufacturing Co., Mt. Pleasant, TX) separating the horses in this small herd, can prevent horses from reaching over and biting one another.

7.4 SHORT PANELS
These standard 56-inch-high (1.4-m) panels are too short for this 16-hand horse. The horse can easily put his head over the panels and get in trouble by nipping at horses being led next to the pen or by chewing on the horse, or the horse's blanket, in the adjoining pen.

The Good, the Bad, and the Ugly

7.5 ALLEYWAY

An alleyway 6 to 8 feet (1.8 to 2.4 m) wide between pens prevents horses from fighting and playing with each other over the panels. A wide alley like this one, along with the use of 6-foot-tall (1.8-m) panels, makes it safer to lead a horse between pens without it being bothered by lunging and biting horses in the pens.

7.6 KICKING DAMAGE

Exuberant playing and aggressive kicking can damage portable metal panels used to separate horses. Straight runs of several panels are unstable and tend to lean over. Stabilize pens by making more angles in the pen or by attaching the panels to posts set firmly in the ground.

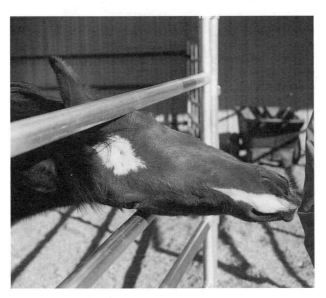

7.7 TOO WIDE

The width between the bars of panels should be appropriate for the size of the horse in the pen. If a horse can put its head between the bars of a panel, it can be a danger to people or horses passing by.

7.8 AN ACCIDENT WAITING TO HAPPEN

If this weanling foal should suddenly startle, he could seriously injure his head.

Connectors

7.9 HINGE-TYPE RODS

Panels like these, with hinge-type rod connectors that hold the panels closely together, are safest for horses. A drawback to these connectors is that the panels are more difficult to set up on ground that is not absolutely flat.

7.10 LOOSE PINS

These connectors are much looser, which makes the panels easy to set up on unlevel terrain. However, the space between the panels is large enough for a horse to get a leg through and become trapped. The smooth curves on the top corners of this type of panel may appear safe, but they can act like a funnel to capture the foot of a rearing horse and leave him hanging. Also, the projecting top portions of the pins are a place for horses to rub or to get a halter or blanket caught.

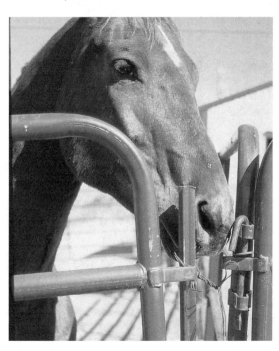

7.11 DANGEROUS JUNCTION

This cobbled-together setup is particularly dangerous. The sharp end of the T-post could cut a horse's face, and the space between the panels is wide enough to catch the front foot of a rearing horse or any hoof of a rolling horse.

7.12 Chain Locks

This type of chain connection (Priefert) is strong, safe, and horseproof, and can be tightened as needed to minimize the space between panels.

7.13 Bolt Clamps

Bolt connectors clamp the panels between two curved steel straps. They make a solid connection and can be used to connect different types of panels, but they provide just the kind of edges horses love to rub on.

7.14 Rubber Straps

Rubber connectors are very quick to connect and disconnect and are suited for situations in which portable panels are needed. The rubber provides an inviting place for horses to rub, however, as evidenced here by the mane hair caught under the connector. A rubbing horse can pop this type of connector loose and escape.

Gate Latches

Gate latches must be impossible for a horse to open and easy for a person to operate, preferably with one hand while wearing winter gloves. No portion of a latch should protrude into a pen where a horse could be injured on it or rub on it. Whether a horse is in the pen or not, gates should be kept latched at all times to keep them from banging.

7.15 CHAIN AND SLOT
One of the simplest and safest latches is the chain and slot. The end of the chain should *always* be secured with a double-ended snap to ensure the gate stays closed should a horse manage to remove the chain from the slot. It takes two hands to pass the chain around the panel to fasten it.

7.16 KIWI LATCH
The Kiwi latch (Colorado Kiwi Co., Steamboat Springs, CO) is horseproof yet easily opened and closed with one hand. This latch doesn't hold a gate solidly in place, but allows it to swing slightly.

7.17 SURE LATCH
This Sure Latch (Co-Line Welding, Inc., Sully, IA) is horseproof and latches automatically. Best suited for a wooden gatepost, it can be secured with a padlock.

7.18 TWO-WAY LATCH
The two-way Sure Latch allows the gate to swing both ways and is easy to operate using one hand.

7.19 SAFETY CHAIN REQUIRED
When a horse leans or rubs on the gate, this sliding bolt latch pops open.

Pen Maintenance

PANEL STAKES

A pushing and rubbing horse can move panels, narrowing the space between pens and allowing the horse to reach objects and other horses that he shouldn't. Anchor pins made from ½-inch-diameter (12.7-mm) steel rods approximately 27 inches (68.6 cm) bent into a U shape keep panels in place. Drive the anchor pin into the ground at an angle because it will hold more securely than one driven vertically. This technique also makes it easier to avoid hitting the lower panel rail with the hammer.

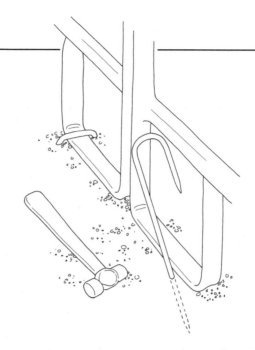

7.20 PERMANENT PEN

A permanent pen welded from steel pipe is virtually indestructible and requires no maintenance. There are no projecting connectors for a horse to rub on. A solid pen is very unforgiving, however, and is more likely to cause injury than a more movable, lightweight pen if a horse collides with it or kicks it.

7.21 TRAFFIC PATTERNS

Obstacles such as large rocks, railroad ties, or even a water tub can be useful pen-management tools. To prevent a horse from rubbing at a particular place, such as at an angled corner, locate obstacles near the base of the panels. If a horse can't get close to a panel, he can't rub. To discourage habitual pawing, place obstacles over a horse's favorite pawing area. Strategically placed obstacles interrupt a horse's traffic pattern and keep him from churning manure until you have a chance to clean the pen.

Pen Maintenance (continued)

7.22 RAILROAD TIE BASE

A horse can relocate panels when he pushes underneath to get grass. Railroad ties, placed against the outside of the pen, can block the space and keep the horse's head inside the pen where it belongs. In addition, the railroad ties help prevent a horse's legs from getting caught under the panels if he happens to roll or lie down near them. Ties are usually heavy enough to stay put, but if a horse consistently moves them, the ties can be anchored to the ground with steel pins.

7.23 PREVENTING RUTS AND HOLES

Horses often create holes by pawing at one particular spot, usually near a gate or along the edge of a pen next to another horse. Rubber mats or used conveyor belting buried 1 to 2 inches (2.5 to 5.1 cm) deep at the horse's favorite pawing area will prevent deep holes from forming.

Warning: Conveyor belting is dangerously slippery when wet, icy, or covered with snow, and it is not recommended for aboveground use where people or horses will be walking.

7.24 LEVEL AND RESURFACE PENS

One advantage of using portable metal panels for pens is that they can be taken down so the ground can be periodically leveled. Once the ground in the pen area is level, fresh gravel is added to maintain a depth of 1 to 2 inches (2.5 to 5.1 cm). Water drains through the gravel and runs out of the pen on top of the base. A tractor blade can be used to level the gravel; the weight of the tractor packs the soil. A slope away from the barn of about 1 inch (2.5 cm) in 3 feet (0.9 m) allows water to drain out of the pens without causing erosion.

7.25 PEN GRAVEL

Washed gravel ⅜ inch (9.7 mm) in diameter or smaller provides some cushion for horses when they lie down. Gravel this size is easy to sift through the tines of a fork when cleaning pens, so minimal gravel is lost with the manure. Larger gravel tends to get lodged in horses' shoes.

7.26 REINFORCE HIGH-TRAFFIC AREAS

Special plastic grids (see Stall Flooring, page 14) can be used to prevent holes from forming in high-traffic areas. In this pen, for example, the resident horse, a vigorous gelding, steps over the railroad tie threshold from his matted eating area onto the gravel at exactly the same spot every time, often at a trot. A deep crater developed. Each time the hole was filled with dirt and gravel it would reappear in a day or two. In addition, the horse's constant movement carried gravel onto the mat in his eating area, which then required frequent sweeping.

The gravel was lowered by 1 inch (2.5 cm; the thickness of the tiles) in a 3-foot-wide (0.9-m) strip in front of the threshold. Tiles were installed and covered with gravel. This prevented holes from forming and reduced the amount of gravel the horse tracked onto his rubber-matted eating area.

Mud Control

7.27 MINIMIZE MUD

Contrary to the old wives' tale, prolonged exposure to water and mud is *not* good for horses' feet. Excess moisture can weaken the hoof and lead to many hoof and skin problems, such as hoof cracks, thrush, scratches, and white line infections.

7.28 DRAINAGE TRENCH

Even well-maintained pens develop depressions that turn into puddles or small ponds. A small trench or furrow cut with a hoe or shovel will allow the puddle to drain, minimizing mud and allowing the ground to dry much more quickly when the sun returns. In areas where the ground freezes, have trenches established before winter.

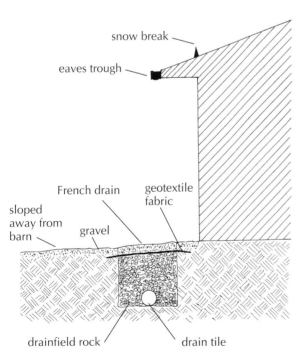

snow break

eaves trough

French drain

geotextile fabric

sloped away from barn

gravel

drainfield rock

drain tile

ROOF DRAINAGE

To help prevent water from coming off the roof and flooding pens, install a gutter system along the edge of the roof. A sawtooth "snow break" attached to the roof above the eaves can minimize snow and ice damage to the gutters by preventing large masses of snow from sliding off the roof.

An alternative is to put a French drain beneath the eaves to carry water away as it falls from the roof. This is a trench 2 feet (0.6 m) wide by 2 feet (0.6 m) deep, with a drain tile in the bottom that carries water to a drainage area. The trench is filled with 2-inch (5.1-cm) drainfield rock to within 6 inches (15.2 cm) of the top. A layer of permeable geotextile fabric prevents the rock from being plugged by dirt and manure, and a 6-inch (15.2-cm) layer of gravel tops it off.

8

✦ VICES AND PREVENTION ✦

Horses have been around for 60 million years but domesticated for only a tiny fraction of that time. That's why they still have strong innate behavior patterns that compel them to bond with other horses, roam and graze continuously, flee from danger, and paw and roll. Take these natural behaviors into consideration as you design your stable management plan.

Vices

In order for a horse to adapt to domestication, his natural behavior must be altered somewhat. Certain horses have a predisposition to neurotic breakdown when faced with domestication pressures such as confinement, domestic feeding practices, and stress. These horses often form vices. *Vices* are undesirable habits horses exhibit in the stable environment that don't directly involve people. Most vices can be prevented. Some vices are incurable.

Prevention

Once certain habits are established, they can be extremely difficult to change, so every effort should be made to prevent a vice from taking hold in the first place. Pawing, for example, is one of the most common vices and is second only to wood chewing in destructiveness. A barefoot horse can wear his hoof so short and imbalanced as to cause lameness, and a shod horse can wear shoes out prematurely. Holes pawed in pens and stalls are dangerous, hold unsanitary moisture, and make cleaning difficult (see chapter 7 for examples of and solutions for pawing).

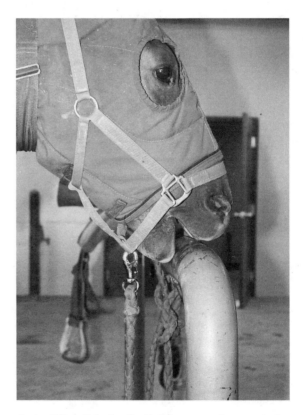

8.1 The Masked Cribber sucks again! Cribbing is not wood chewing. It's when a horse clamps onto a solid object — any object, from a pipe rail to a water bucket — tenses his neck muscles, and swallows air. This horse wears a cribbing strap unless he is being ridden. The strap was removed and the horse immediately started cribbing at the hitch rail.

Common Vices

Vice	Description	Causes	Treatment
Cribbing	Anchoring of incisors on object (post, stall ledge), arching neck, gulping air; can cause colic and poor condition	Initially boredom, imbalanced diet, internal parasites; it's thought that cribbing releases endorphins that stimulate pleasure center of brain, making cribbing addictive	Can be incurable; cribbing straps, shocking devices, and muzzles have all shown results with some horses; future drug treatments possible; surgery possible
Pawing	Digging holes; tipping over feeders and waterers; getting leg caught in fence; wearing hooves and shoes excessively	Initially confinement; boredom; anxiety; excess feed; lack of exercise	Curable; increase exercise, diversions; don't use ground feeders and waterers; don't reinforce by feeding; formal restraint lessons; use rubber mats (see chapter 7)
Self-mutilation	Biting flanks, front legs, chest, scrotal area with squealing, pawing, and kicking out; onset 2 years, primarily stallions	Can be endorphin addiction similar to cribbing; can be triggered by confinement, lack of exercise, sexual frustration, hormones	Manageable, might be curable; geld nonbreeding stallions; increase exercise; provide larger confinement area, stall companion, or toys; neck cradle; muzzle; future pharmacological treatment possible
Stall kicking	Kicking stall walls and doors with hind hooves, resulting in facilities damage, hoof and leg injuries, lost shoes	Confinement; boredom; some horses seem to like the sound; irritation at neighbor; rewarded by attention; rodents in stalls; itchy heels; mimicking	Can be curable depending on how long-standing the habit; increase exercise; pad stall walls or hooves; kicking chains useful in some cases; don't reinforce by feeding
Tail and mane rubbing	Rhythmically rubbing the rear end or base of the mane against an edge or surface such as a fence or stall wall	Initially dirty udder, sheath, mane, or tail; shedding hair on hindquarters; pinworms, ticks and other external parasites; skin infections; allergies to fly repellents; shampoo not rinsed out	Manageable; groom thoroughly and often; clean sheath and udder; deworm regularly; eliminate hard edges and protrusions in pens and stalls; install electric fence
Weaving/pacing	Swaying back and forth, often by doors or gates; repeatedly walking a path back and forth	Confinement, boredom, excess feed, high-strung or stressed horse, lack of exercise	Manageable; provide as much exercise as possible; turn out where he can see other horses; use specially fitted stall door for weaver; provide stall toys for diversion; use doubled hay net to extend feeding time
Wood chewing	Gnawing of wood fences, feeders, stall walls; can ingest up to 3 pounds (1.4 kg) of wood per day; socially contagious	Diet deficient in coarse roughage and/or minerals; boredom; teething; stress habit; stabled with other wood chewers	Manageable; increase roughage in diet; apply antichew products to decrease palatability of wood; increase exercise, time out on pasture; provide free-choice salt and minerals

Wood Chewing

Wood chewing can afflict a horse of any age and can result in colic from wood ingestion or damage to the gums and lips from splinters, to say nothing of the damage to facilities. Young horses may begin nibbling out of boredom, curiosity, or as a response to teething. A horse that lacks sufficient salt in its diet will sometimes chew on any surface that presents itself. Serious woodchewing can initially be caused by low fiber intake in relation to a horse's needs, especially during cold or wet weather. Weather-related wood chewing is thought to be a result of the frustration and anxiety felt when the animal is uncomfortable. Water also softens the wood, making it more palatable and aromatic. Once a horse has developed the habit of wood chewing, he will often continue it, and may chew rubber, plastic objects, and metal even though the initial cause of wood chewing no longer exists.

Be sure a horse always has access to free-choice salts and receives adequate fiber in the form of long-stem hay. During cold or wet weather, increase the roughage portion of the horse's ration.

8.2 The beaverlike gnawing of wood planks, rails, buildings, trees, and feeders is unhealthy, costly, and unnecessary.

8.3 Modern electric fence wire, made of plastic with interwoven stainless steel wires, is inexpensive and easy to use. Wide versions are called tape or ribbon. Here, one electric wire protects the horizontal fence brace, and another prevents horses from reaching through the fence.

DISCOURAGING CHEWERS

There are many commercial products (see appendix) designed to make wood unpalatable to horses. Some contain substances like hot pepper or bitter apple, but the most effective ones resemble creosote. Oily products can be messy. Many antichew products are available in clear formulas for protecting natural-colored wood surfaces such as doors and walls. A plastic squeeze bottle, such as from mustard, ketchup, or shampoo, can make an efficient applicator. Use a paintbrush to spread the product evenly on the wood.

Some horses chew anything they can get their teeth onto, including metal gates and plastic electric fence insulators. Most antichew products are ineffective on non-porous surfaces such as metal and plastic. Paste products can be used in such instances, but be aware that a horse that rubs on the treated surface will get the paste on his head and coat. A cleaner method that's worked in some cases is to rub the surface you want to protect with a bar of strongly scented soap.

Wood Chewing

8.4 MUZZLE

A well-fitted muzzle is a very low-maintenance, cost-effective way to prevent chewing (check tack stores for availability; see Appendix). Check muzzles regularly to ensure proper fit. Make sure there is no place in the turnout area where a horse can catch the muzzle, and even then, attach a muzzle only to a breakaway halter.

8.5 METAL EDGING

One of the best antichewing preventives is to apply metal corners to wood edges. The metal should be at least 0.05 inch thick (1.3 mm; 18 gauge) and fastened securely at frequent intervals. You can make your own, or metal shops can custom-bend sheet metal. Manufactured corners in galvanized and painted finishes are available (see Appendix).

HOW TO PREVENT VICES

- Educate yourself so you understand horse behavior, horse needs, and the causes of vices.

- Be sure the horse receives thorough ground training.

- Provide every horse with active, sustained exercise at least five times a week.

- Allow every horse plenty of turnout time to "just be a horse."

- Tailor each horse's ration to meet his energy needs; do not overfeed.

- Provide a stall toy to occupy a bored horse.

- Use stall grilles and/or windows that open to allow a horse to see other horses and watch barn activity.

- Whenever practical, immediately reprimand a horse, by a slap and a verbal command such as "Stop," for any undesired behavior that might lead to a vice.

- Do not inadvertently reward a horse for behavior that might lead to a vice, such as feeding a horse that is pawing or kicking the stall.

- Modify facilities to discourage vices such as chewing, rubbing, and pacing.

A

B

8.6 Nip Chewing in the Bud

The sill of this stall doorway (A) is a typical example of the beginning stage of serious wood chewing damage. In this instance, repeated applications of a variety of antichew products failed to stop the horse from chewing. The problem was solved by fitting a piece of galvanized metal to the sill and attaching it with screws (B).

8.7 Don't Use Drywall Corners

Even though the preformed metal corners used to finish drywall or sheet rock are readily available, they are far too lightweight to be used for chewing protection. As seen here, horses can easily rip the thin metal, resulting in dangerous sharp edges. Also, a horse could be seriously injured by ingesting metal shreds.

8.8 Protect Trees

A tree will die if a horse removes the bark. Be sure horses with access to trees get plenty of hay and have a free-choice trace mineral block. Though some antichew products can safely be used on trees, they require frequent application. More positive protection is provided by a fence that keeps horses at a distance. A pipe fence, as shown here, is better than a wood fence, which itself might need to be treated to prevent chewing. A single electric fence wire, if well maintained, is a simple and effective solution that is easier to install but requires a reliable power source.

Alternatively, trees can be wrapped with heavy wire or chicken wire. Tie the ends of the wire securely to keep them from coming loose and injuring your horse. Some horses make a wire-covered tree their favorite rubbing spot. Monitor tree growth and adjust or replace the wire, if necessary, so it does not become embedded in the bark.

9

✦ SANITATION & PEST CONTROL ✦

Healthy, sanitary conditions in a stable are a direct result of good management practices. These include proper disposal of manure and other wastes, control of rodents, birds, flies and other insects, and keeping moisture in the barn area to a minimum (see chapter 7 for tips on mud control). Effective sanitation practices must be implemented for your horse's health, your family's health, your relations with your neighbors, and to satisfy your legal obligations.

A 1,000-pound (454-kg) horse produces approximately 50 pounds (22.7 kg) of manure and 6 gallons (22.7 L) of urine every day. Manure must be properly managed to control odor, remove insect breeding areas, and kill parasite eggs and larvae. Urine contains urea and hippuric acid, both of which produce ammonia as they break down. Especially in an enclosed barn, pungent odors can create an unpleasant work environment, and ammonia vapor can be injurious to the eyes and lungs of both humans and horses. It also destroys tack by drawing out the oils in the leather. The combination of dung and urine is a perfect medium for the proliferation of bacteria that can break down a horse's hooves. Flies are attracted to odors such as manure and wet bedding, which is where they like to lay their eggs.

Bedding

Bedding provides a comfortable surface for a horse to lie on and absorbs urine. Bedding insulates a horse from the cold and damp of the floor and protects his fetlocks, hocks, and hips from abrasion. A stall should be bedded to invite a horse to lie down, so he can rest his legs and prevent "stocking up" or swelling of the lower legs. Bedding material should be absorbent, soft, dust-free, easy to handle, and nontoxic. Absorbent bedding facilitates the removal of urine. The type of bedding you use will depend largely upon what is available in your area, but be prepared to switch bedding, because horses can show allergic reactions to certain types of sawdust, shredded paper, and dusty straw. Although sand makes a cool, comfortable bedding, it is undesirable because it encourages pawing and increases the risk of colic. Some horses intentionally eat sand (and soil), especially if not fed sufficient salt, and other horses inadvertently ingest sand along with their hay.

How much bedding you use, that is, how deeply you bed the stalls, depends on the horse's needs and habits and your personal preference. A comfortable bed for most horses would be provided by 4 inches (10.2 cm) of sawdust or wood chips, 6 inches (15.2 cm) of shavings or paper, or 12 inches (30.5 cm) of straw. Less bedding makes a stall easier to pick out, but can lead to abrasions on a horse's bony prominences from insufficient cushion. Deeper bedding is more difficult to clean, especially if not cleaned frequently, but might be necessary for thin-skinned horses.

Straw

Straw is traditional horse bedding (photo 9.1) and is readily available in farm country. When composted, straw decomposes the fastest of all bedding materials. It can be quite dusty, however, which can cause respiratory problems, and it doesn't absorb odors or moisture nearly as well as other types of bedding. Wet straw is slippery underfoot and is easily tracked from the stall into the barn aisle. When cleaning a stall, it's difficult to separate clean straw from manure and wet straw, and much good bedding is often thrown out with the bad. Digestive problems can result from a horse eating certain types of straw, such as oat and barley. If you use high quality straw to bed the stall of a chronic overeater, you'll likely find a bare stall in the morning.

Wood Products

Wood products (shavings, sawdust, and wood chips) are the most common type of horse bedding, and are the easiest to handle. They are available bulk in timber country, and packaged in most other areas. Chips, shavings, and sawdust vary in their absorbency, loft, and dust level. Shaving and chips won't ball up in a horse's feet as much as sawdust will. Fresh or "green" products are less absorbent than dry ones, and hardwoods are generally not as absorbent as softwoods.

Walnut products, which tend to be dark in color, are considered toxic to horses, and can cause colic and laminitis within hours of contact. Furthermore, some horses have shown adverse reactions to oak and hickory and other types of wood bedding. In some cases, a horse's feet may become too dry when bedded on sawdust, particularly oak, for long periods.

With bulk products, watch for foreign objects such as wood splinters, nails, and wire. Baled or packaged bedding is easier to handle than bulk bedding, and is cleaner and of a more consistent quality.

Shredded Newsprint

Shredded newsprint, as from newspapers and phone books (not magazines, computer paper, or colored pages), is the most absorbent bedding but also the most flammable. All U.S. inks are soy-based and contain no lead, and horses are not harmed by ingesting newspaper. Newspaper bedding turns a dirty gray color in stalls, which is disagreeable to some people, and the ink can stain light-colored horses. Paper is as warm as sawdust and straw if it remains a uniform depth, but it has a

9.1 Fresh straw makes one of the softest, most inviting beds for a horse to lie on.

9.2 Shredded paper usually doesn't stick in the manes and tails and is virtually dust-free.

tendency to ball up in the corners of the stall as the horse moves around.

Pieces of shredded paper should be no smaller than 1 by 1½ inches (2.5 by 3.8 cm), to prevent inhalation. Pieces longer than 5 inches (12.7 cm) become entangled in horses' feet and get dragged out of the stall. Long strips don't absorb as much moisture as smaller pieces, and are more likely to be blown around the barn and pastures after the stalls are cleaned. Wet paper compacts into a very hard layer that makes cleaning the stall difficult.

Newsprint contains organic matter that makes it equal or superior to other kinds of bedding in fertilizing soils; it decomposes at a similar rate to straw. However, newsprint that has not been thoroughly composted blows about when handled, tangles in the spreader, and when spread on fields is not visually pleasing.

Sources include recycling centers and newspaper publishers, or contact your county's recycling coordinator or county extension agent.

Cleaning a Stall

Stalls should be cleaned at least twice a day, more often if possible. Dry footing is essential to a horse's overall health and particularly to the soundness and condition of his feet. Simply adding fresh bedding and allowing manure and soiled bedding to accumulate in the horse stall results in dirty animals and blankets, increased odors, unhealthy air, constant exposure to parasite infestation, and an ideal fly-breeding environment. It's much easier and faster to clean a stall twice a day than to overhaul one that's been let go for a day or more. Bedding use can be kept to a minimum if sufficient time is taken to separate clean bedding from soiled bedding, and the more regularly a stall is cleaned, the easier this is to do.

A stall can be cleaned more thoroughly and safely when the horse is turned out. When cleaning an occupied stall, tie the horse safely to keep him from charging the door, and never leave a cart, fork, or rake unattended or where the horse can get it. Be careful of touching the horse on the legs with the cleaning tools, which might cause him to kick.

The easier it is to clean the stalls the more likely you'll be to clean them regularly. To make the cleaning process as convenient as possible, you'll need to prepare the stall properly (see Stall Flooring, chapter 2), choose the right tools for the job (see chapter 6), and find a good way to dispose of the used bedding.

TYPES OF BEDDING

BEDDING	ABSORBENCY	COMFORT	CONVENIENCE	COMMENTS
Straw	Fair to poor	Good when fresh and deep	Light bales	Dusty; can cause colic when ingested
Pine sawdust	Good	Warm, soft	Bulk or bags	Dusty; nice aroma; can be drying to hooves
Pine shavings	Fair	Warm, fluffy	Bulk, bags, or bales	Nice aroma
Shredded newspaper	Excellent	Warm and fluffy when fresh; dust-free	Bags or bales, limited availability	Difficult to clean; messy; flammable
Hardwood chips, shavings, etc.	Poor	Can be rough	Bulk	Some are toxic to horses; can be drying to hooves; not recommended
Sand	Very poor	Soft but abrasive if not deep	Bulk	High risk of colic; not recommended

Cleaning a Stall

9.3 PICK UP MANURE
Pick up manure and wet bedding with a fork, removing as little clean bedding as possible. Fine bedding, such as sawdust, is easier to sift through a fork than shavings or straw. Separate the cleanest bedding and move it against the walls.

9.4 SCOOP WET SPOTS
Use a shovel, such as an aluminum scoop shovel, to scoop up packed or urine-saturated bedding that can't be picked up with the fork.

9.5 SWEEP BEDDING
Use a push broom to sweep a small amount of the oldest dry bedding back and forth over wet areas to soak up any remaining liquid, then pick the bedding up with the shovel. The floor should be absolutely clean and as dry as you can get it.

9.6 DRY THE FLOOR
If possible, leave the floor uncovered to air-dry. If the stall is to be used again immediately, sprinkle an odor neutralizing product (see chart) over the wet area and cover it with the oldest clean bedding left in the stall. The cleanest bedding along the walls and any new bedding that's added should go where the horse lies down. Bank some bedding along the walls to minimize the chance of a horse getting cast (when a horse rolls over against the wall and can't get back up). Leave the floor bare where the horse is fed, beneath the water bucket, and in front of the door, to minimize bedding being tracked into the aisle.

Odor Control Products

When stalls are used regularly, they often can't dry out enough to get rid of odors. Hydrated lime was the standard for many years for deodorizing and drying out stall floors. Lime is strongly alkaline, however, and can be very harmful to people and horses. Baking soda (sodium bicarbonate) is safer and more effective than lime but it is not very absorbent, and its fine consistency requires quite a bit to be used in a stall, which makes it less cost-effective than other products. Products made from zeolite (a class of inorganic oxide materials) are nontoxic to people and animals, whether they are ingested, inhaled, on the skin, or in the eyes. They are nonflammable, environmentally friendly, and can be safely used around horses' water and feed. Zeolites are safe enough to be handled with bare hands, which is especially nice for those suffering from chemical sensitivity.

STALL DEODORIZERS

PRODUCT NAME (active ingredient)	APPROXIMATE COVERAGE* Weight	No. of stalls	APPLICATION METHOD	PROS	CONS
Sweet PDZ Stall Refresher (zeolite) [a]	40 lb (18.1 kg)	15	Sprinkle directly on wet bedding or stall floor	Very safe, nontoxic, very absorbent, excellent deodorizer, more than 3,000 dealers	None
Odor Capture (cotton lint and pecan pith) [b]	30 lb (13.6 kg)	1	Sprinkle directly on wet bedding or stall floor	Very safe, nontoxic, absorbent, excellent deodorizer	Faint sour odor
Stable Boy (zeolite) [c]	44 lb (20 kg)	15	Sprinkle directly on wet bedding or stall floor	Very safe, nontoxic, very absorbent, excellent deodorizer	None
Baking soda (sodium bicarbonate)	50 lb (22.7 kg)	15	Remove soiled bedding, sprinkle on wet floor areas, allow to dry, add fresh bedding	Very safe, good deodorizer, readily available	Not very absorbent
Stall Odor Kleen (potassium dichromate) [d]	16 oz (0.4 L)	32	Remove soiled bedding, spray on wet floor areas, allow to dry, add fresh bedding	Most economical product per use, excellent deodorizer	Toxic, danger to eyes, skin, and mucous membranes; doesn't help stall dry out
Hydrated lime (calcium hydroxide)	50 lb (22.7 kg)	20	Remove soiled bedding, sprinkle on wet floor areas, allow to dry, remove lime and add fresh bedding	Inexpensive, readily available	Very alkaline, can irritate skin, mouth, throat, lungs; not very absorbent; slippery when wet

*Coverage estimates are based on a 10' x 12' (3 x 3.7 m) stall.

[a] Steelhead Specialty Minerals, Inc. (Spokane, WA)
[b] Advanced Environmental Solutions (Phoenix, AZ)
[c] Westhawk Traders (Vancouver, BC, Canada)
[d] G. G. Bean, Inc. (Brunswick, ME)

Disposal

Horse manure is considered one of the best natural fertilizers because it is quite high in nitrogen and contains valuable organic matter and trace minerals. Although few problems are encountered with applying fresh horse manure to established grass pastures, it should never be applied to a garden or to newly planted trees since it is likely to "burn" plant tissues.

Once manure is collected, it can be hauled away, spread immediately on a pasture or trails, or composted on a pile for later distribution. Composting is the most common method and is convenient, because manure doesn't have to be hauled every day. If you compost manure, it reduces bulk and concentrates soil nutrients. The end product of composting is humus, an odorless, finely textured material that makes an excellent soil conditioner and fertilizer. Local mushroom farmers and gardeners are often on the lookout for well-composted horse manure and will often load it and haul it away for free. Other outlets for compost are greenhouses, nurseries, botanical parks, and topsoil companies.

Composting

Composting, the transformation by bacteria of manure and other organic materials into humus, a dark, pleasant-smelling soil, can take anywhere from two weeks to several months. Keeping the pile moist (not wet) and mixing it frequently will speed up the process, kill parasite eggs and larvae, and reduce the fly population. Most organic materials can be safely composted, but avoid adding the following to the pile:

- Cat and dog waste
- Meat, fish, fat, and dairy products
- Diseased or insect-infested plants
- Anything treated with herbicides or pesticides
- Weeds with mature seeds
- Rhubarb and walnut leaves

9.7 Especially important for urban acreages, some refuse collection businesses offer a specialized manure hauling service, while others are willing to haul it away along with normal trash. This farm has a 40-cubic-foot (1.1-m^3) Dumpster that is emptied twice a week.

9.8 Composted horse manure is a fertilizer prized by gardeners, mushroom growers, and greenhouse operators. Even if there's not a cash market for it in your area, with a little advertising you're likely to find folks who will load it and haul it away for free.

USING A SPREADER

If you're going to spread the manure right away, it makes sense to clean stalls or pens directly into the spreader. It is not a good idea, however, to store manure in a spreader for any length of time, because it will soon rot the wooden floor and rust the metal sides.

Building a Manure Pile

9.9 INEFFICIENT PILES

An untidy compost heap is not only unsightly, but inefficient. Depositing manure in many small piles wastes space and is not conducive to effective composting.

9.10 ORGANIZED PILE

Use the first load as a base, and dump each succeeding cartload on top of it. Locating the compost pile on a slope makes pushing the full cart to the pile easier; as the pile grows, the top of the pile becomes level. Use the cart to level the pile and form a smooth delivery ramp.

9.11 MANURE FORT

If you have a hillside or bank near the barn, a bunker or "manure fort" is a handy way to manage your compost pile. This fort was cut into a bank and lined on three sides with railroad ties, stacked so the sides of the bunker slope outward. Initially, manure is simply dumped over the edge. When the manure reaches the top of the fort, you can walk onto the pile to dump the cart, extending the level pile as far as needed.

9.12 SPREADING MANURE

When it's time to spread the manure it can be loaded into a spreader by hand, but a tractor with a loader makes the job go faster. In winter climates, the best time to spread manure is in late fall or early winter, when the pastures are dry or frozen and free of snow. Ensuing snow cover will keep the compost from blowing away and encourage the nutrients to soak into the ground.

Parasite Control

All horses have worms, or parasites, and wherever there is manure, there are parasite larvae. The life cycle of all horse parasites involves leaving the horse host via the manure and then infesting a new host (or the same host again) when a horse ingests parasite eggs from manure-contaminated ground. Parasite larvae, especially strongyles, can do great internal damage to a horse as they migrate through the tissues, and can cause thromboembolic colic. Keep parasites under control with good sanitation, manure removal, bot egg removal, and deworming.

Your deworming program should target strongyles (bloodworms), ascarids (roundworms), *Oxyuris equi* (pinworms), and *Gasterophilus* (bots). Adult horses should be dewormed every two months year-round, unless on a daily feed-through program. Foals should be dewormed every month, from one month of age until they are weaned, then every six weeks until they are one year old. After that they can join the adult program of six times a year. At least twice a year, use a product such as ivermectin that is effective against all worms and bots. Feed-through dewormers added on a daily basis to the horse's normal grain ration are especially effective for situations with high horse density. Feed-through dewormers can be used for continuous control of strongyles, bloodworms, ascarids, and pinworms. They do not remove bot larvae.

- ▶ **Spring** (April or May, depending on your climate): Use a bot dewormer to kill bot larvae before they leave the horse's stomach.
- ▶ **Fall** (October or November, depending on your climate): Use a bot dewormer after a killing frost, and after all bot eggs have been removed from the horse's coat.
- ▶ **Twice in summer and twice in winter:** Use ivermectin, or choose from other dewormers, paying close attention to their effectiveness against strongyles, which are the biggest parasite threat to your horse's health.

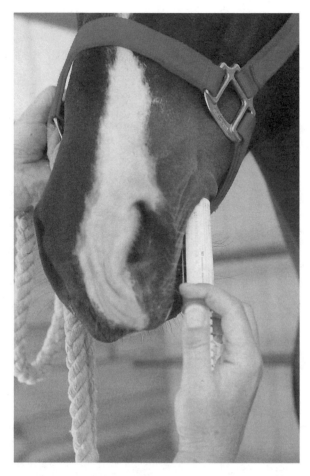

9.13 Paste dewormers are commonly sold in plastic syringes calibrated in body-weight increments that make it handy to inject the proper dosage into the horse's mouth. If the horse's mouth is clean, the dewormer will be distributed on his tongue and the roof of his mouth and he'll likely swallow the dewormer at once. If there are wads of hay or grass hidden in the horse's cheeks, however, the horse can spit the whole kit and kaboodle out onto the ground or your shoe.

Parasite Control

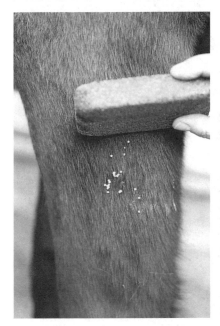

9.14 USING A BOT BLOCK
Use a bot block or other means to remove the sticky eggs of bot flies as soon as you spot them.

9.15 SHARPENING EDGE
To clean a bot block and "sharpen" it, draw the surface of the block along a sharp edge, such as the end of a board.

9.16 SHARP EDGE
The hair and dirt-filled material are scraped off the edge of the block, leaving a clean surface with open pores.

METHODS OF BOT EGG REMOVAL

There are several ways to remove bot eggs:

▶ Wash the legs with hot soapy water or place a warm cloth on the eggs. This will hatch the eggs and the larvae will die, eliminating the possibility of the horse ingesting them.

▶ "Sand" the eggs off with an abrasive pad designed for nonstick cookware, sandpaper (fine to medium), or a special cinder block, like that used to clean grills.

▶ Coat the affected areas with baby oil, vegetable oil, or petroleum jelly to loosen the eggs and make them easier to scrape off.

▶ Scrape the eggs off using a special bot knife, a disposable razor, or a dull hacksaw blade.

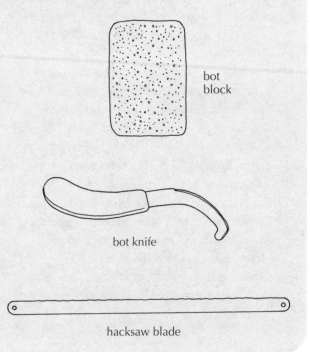

bot block

bot knife

hacksaw blade

Fly Control

Flies spread diseases by carrying blood or mucus from one animal to another. Pesky flies can interrupt feeding and cause constant evasive movements, leading to weight loss and loose shoes. Flies can also make a horse fidgety during grooming and drive your horse to distraction when you're riding. Flies feed and reproduce on filth, decaying matter, and water, and in one season can breed as many as twenty-five generations. Good sanitation and management practices are the first step in reducing fly populations.

Fly Bait

Poison fly bait is an insecticide combined with a sex attractant and fly-feeding stimulants that attract both sexes of the housefly and many other species of filth flies, including blowflies and eye gnats. It does not attract biting flies such as horn flies and stable flies. The bait attracts flies in a 250-square-foot (23.2-m^2) area. Once exposed to the air, the attractant in the bait dissipates. Depending on the environment, the bait remains effective from several hours to several days before fresh bait must be added. Some people find the bait odor offensive, and the insecticide is harmful to humans and animals if swallowed, inhaled, or absorbed through the skin. The bait should be inaccessible to animals and children, which limits its use around horse barns.

Fly Parasitoids

Fly parasitoids, or "predators," are one-eighth the size of a fly and don't bite, sting, crawl on people or animals, or migrate into your house (photo 9.18). What they do is spend their entire life seeking out fly pupae, in manure or rotting vegetation, in which to lay their eggs. Fly parasitoids won't kill adult flies but they will kill the pupae of common manure and filth-breeding pest flies, including house flies, biting stable flies, horn flies, garbage flies, and blow flies.

Fly parasitoids are shipped in the immature stage packed in wood shavings. They are applied by sprinkling the shavings over areas where manure and urine are accumulated and places where flies are crawling and breeding. They hatch and take off from there, ranging up to 80 yards (73.1 m).

9.17 Cleaning up old hay around pens and feeders eliminates fly breeding areas and is the first step in controlling flies.

FLY CONTROL OPTIONS

- Eliminate breeding areas
- Fly predators
- Sticky paper
- Sticky tape
- Sticky tube
- Traps
- Fly spray

- Fly bait
- Fly cream
- Fly strips
- Fly masks
- Fly sheets
- Feed-through larvicide
- Poison misting systems

9.18 **Fly parasitoid on a fly pupa**

Fly parasitoids work best if you start early in the season, before pest flies mature, and stay with the program, replenishing the parasitoids monthly during the fly season. Parasitoids are vulnerable to insecticides, so when applying fly spray be careful that it doesn't drift onto manure piles or other areas containing parasitoids.

Fly Papers and Tapes

Flypaper is a sticky ribbon of paper, usually 2 inches (5.1 cm) wide, that holds flies and any other insects, or even small birds, when they contact it. It comes rolled in a cardboard tube and is usually hung by a string using a thumbtack. (If you lose the thumbtack in your horse's bedding, comb the area with a magnet to find it.) Flypaper is biodegradable and contains no chemical attractants or poisons. It has a shelf life of two years and need not be protected from freezing. The more dust in the air, the quicker flypaper loses its stickiness. If flies are landing and leaving, it's time to replace the flypaper.

The sticky fly tape system (photo 9.19) consists of 1,000 feet (305 m) of sticky white polypropylene tape on a reel. Using plastic pulleys, the tape can be run the length of the barn, or in any desired configuration on a wall or ceiling. It's not affected by freezing, so you can leave it up all winter, but dust will likely cover the tape, so you'll need to roll out fresh tape in the spring.

Flies are especially fond of narrow places, such as garbage can rims and electric wires, which is why the narrow white tape is so effective (photo 9.20). Like flypaper, it contains no chemical attractants or insecticides, is nontoxic, and has a shelf life of five years.

A different form of sticky trap is in the shape of a tube (photo 9.21). It's about twice as effective as ordinary flypaper because it uses a granulated attractant to draw flies to the sticky surface. Although the trap is nontoxic, it's extremely sticky, and it should be located out of reach of horses, children, and pets, and where birds and bats won't fly into it.

The sticky substance used on most sticky traps can be removed from hair and skin with baby, mineral, or vegetable oil.

Water and Bag-type Traps

Stinky water traps attract houseflies and other filth flies, but not biting flies (photo 9.22). Flies must be flying and feeding for water traps to work. Commercial bait contains various sex pheromones and feeding stimulants that smell like the things flies like to eat and lay eggs in. A piece of raw meat will work as well. Once dead flies begin decomposing in the jar, they act as an additional attractant.

9.19 Sticky Fly Tape System

9.20 Close-up of fly tape

9.21 Sticky tube

Flies are drawn from more than 20 yards (18.3 m) away; traps should be spaced farther apart than that unless you have a very dense fly population. If fifty to one hundred flies are not caught within 5 days, there either are not enough flies around, it's too cool (flies won't feed or seek food below 70°F [21.1°C]), or the traps are too close together.

For optimum results, water traps should be located on the ground or hung no higher than 4 feet (1.2 m) because fly scent clings to the earth, forming a path that draws flies to the trap. Generally, traps work best if set in the sun, but when the temperature is above 90°F (32.2°C), flies seek shade to cool off. Place traps in the sun at lower temperatures and in the shade when it's hot. Place traps 10 to 25 feet (3 to 7.6 m) from areas you want to draw flies away from, and preferably downwind so you don't have to smell the traps. Add water as needed to keep the liquid level to within ½ inch (12.7 mm) below the surface of the fly mass and agitate the trap frequently to keep maggots (fly larvae) from emerging and odors active. Maggots will lower the trapping ability and attraction of the trap. If maggots emerge, add 2 ounces (47.4 mL) of

FLY LARVICIDE

A feed supplement containing a fly larvicide can be fed daily to prevent the development of stable and house flies in manure (see appendix). This method does not kill adult flies. Manure from horses eating the larvicide is not harmful to animals that contact it. For best results, start feeding early in the spring before flies appear and continue until cold weather ends the fly season.

vegetable oil to the trap, shake it well, let it stand overnight, and then empty the trap and start over with fresh bait.

Bottle traps can be emptied (dead flies are said to be particularly good fertilizer for roses), washed out, and baited with fresh meat or commercial attractant and water, or with a 1-inch (2.5-cm) layer of dead flies saved from the full trap (this is not a job for those with weak stomachs).

9.22 Stinky water traps attract flies by odor.

9.23 Bag-type traps work like bottle traps and are disposable.

Fly Spray

Fly sprays are the final line of defense and should be used as only a small part of an effective fly control program. Fly spray is most appropriate as a temporary means to keep flies off a horse when you're riding and training or when the horse is trying to relax on an especially bad fly day. It's not realistic to expect to provide a horse with 100 percent insect relief from fly spray alone.

Most fly sprays contain insecticides that kill flies, but in order for flies to die they have to land on the sprayed horse and touch or ingest the poison in the fly spray. Few fly sprays are truly successful at repelling flies, that is, keeping them from landing. Some horses and people are sensitive to fly sprays, even those containing "natural" ingredients, and can suffer adverse reactions such as skin rash, breathing difficulties, and nausea.

Spraying can be wasteful and ineffective if not done properly. Never spray the horse's head or face. Besides causing the horse to pull back, most sprays are irritating to eyes and mucous membranes.

First, brush the horse to remove all loose hair, dirt, and debris. If the horse is very dirty, bathe him and make sure the coat is dry before applying fly spray. Brush against the hair coat while spraying, to distribute the product evenly throughout the hair and work it into the coat, where it is protected from being broken down by sunlight.

Fly sprays come either in ready-to-use (RTU) form, usually sold in a spray bottle, or in a concentrated form that is diluted with water before using. Concentrated forms are much more economical to use and can be mixed stronger for tough flies.

Liquid wipe-on products are often oil-based and usually last longer between applications than water-based sprays. But pouring liquid on a cloth or mitt and wiping it on can result in uneven application and wet spots that soak a horse's skin, which many product labels warn against.

Presaturated towelettes are very convenient for trail rides and for applying repellent to a horse's head, belly, and ears.

Some spray products are also labeled as "premise" sprays, for use on barn surfaces and in cracks to kill flies and other insects.

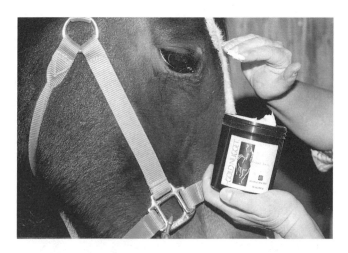

9.24 FLY CREAM
It's difficult to spray the areas on a horse that biting insects find most inviting: the ears, the furrow under the jaw, the creases along the front of the neck, the chest, and the belly just ahead of the sheath or udder. It's easier and more effective to apply a cream to these places.

Some creams, such as Gnat-Away (Neogen Corp., Lexington, KY) can be used on sensitive areas of a horse such as around the eyes, sheath, and udder. It can also be applied sparingly to the inner ear with a long cotton swab. SWAT Fly Repellent Ointment (Farnam Companies, Inc., Phoenix, AZ) is one of the few repellents labeled for application directly to superficial wounds and open sores.

CAUTION
Some people and horses are sensitive to the chemicals in insecticides. Most fly sprays kill beneficial insects as well as flies, and can endanger birds and fish, if overspray or wash water containing the poison enters a stream or pond.

Systems that are timed to spray insecticide into the air at predetermined intervals can virtually eliminate flies in a barn. A pump feeds insecticide from a reservoir to nozzles located throughout the barn.

Fly Control

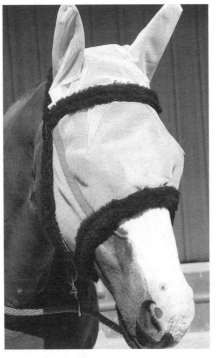

9.25 FLY STRIPS

Plastic strips impregnated with insecticide can be attached to a horse's halter; some come with a special nylon collar that fastens with Velcro. If you turn out a horse that's wearing a halter, make sure the halter is a breakaway style to minimize injury if the horse catches the halter on something.

9.26 FLY MASK

A fly mask can give your horse peace of mind and protect the sensitive areas of his head without the use of chemicals. Masks are especially useful for horses with light-sensitive eye disorders because they block about 70 percent of the sun's harmful rays. A mask can also protect a horse from wind-blown debris while trailering, and can protect an eye that's being treated for an injury. Some masks and bonnets are designed to be used with a bridle while riding.

9.27 FLY SHEET

Fly sheets are made from a mesh fabric designed to keep a horse cool and prevent flies from landing on the horse's body. Some are made from cotton/polyester or nylon, but the toughest sheets (Horseware Triple Crown Blanket, Kinston, NC) are made from the same material used for lawn furniture (woven polyester yarns coated with PVC). These sheets are tough enough for turnout.

Stable Pests

Rodents have no place in a horse stable. They can spoil feed, cause fires by chewing insulation off wiring, ruin tack, and frighten horses into kicking. Birds are less of a danger, but their droppings can make a mess. Some horses and people are allergic to wasp stings, and it's best to eliminate nests as soon as you spot them.

Rodents

Perhaps the best means of rodent control is nature's own mobile mouse catcher: the cat. A well-cared-for neutered cat can easily provide more than 10 years of extermination service as well as companionship for you and your horses.

A few strategically placed, well-monitored mousetraps can drastically reduce the mouse population in the barn. New paddle-style traps are foolproof to set and easy to empty, a big improvement over old fashioned trigger-style traps that caught more fingers than mice. Peanut butter is the best bait.

Birds

Birds reduce the insect population around a barn and they can be a joy to watch as they build nests and raise families, but their droppings can be a constant nuisance. The best way to keep birds out is to apply screens over open windows and keep doors closed, especially during nesting season. Once birds have established nests elsewhere, they are less likely to come into the barn, and the barn can be left open. The ideal situation is to encourage birds to nest *near* the barn but not *in* the barn.

Wasps and Hornets

The most common type of wasp nests found around barns are made by paper wasps, mud daubers, and bald-faced hornets. The best way to get rid of wasps is to use an aerosol insecticide especially formulated for wasps and hornets, and spray at night, when the insects are inactive. Before spraying, remove your horses from the area, wear a hat, jacket, and scarf, and direct the narrow stream of insecticide right into the entrance hole of the nest.

MOUSEPROOF DOORS

To keep mice out of the feed room and tack room it is essential to block the space between the bottom of the door and the floor. One way to do this is with a well-fitted door sill. Sills are available at hardware and building supply stores or can be custom made to fit uneven spaces.

Another way to make a door mouseproof is to attach a rubber strip to the bottom edge, so it just touches the floor along its length, and touches the sides of the jamb when the door is closed.

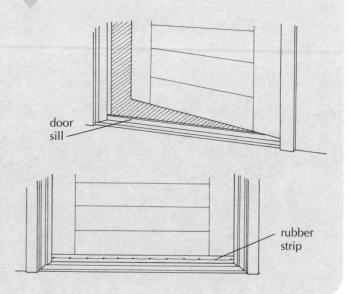

door sill

rubber strip

✦ HAY ✦

Hay is the mainstay of most horses' diets. When fed good hay, horses thrive and require little grain supplementation. Making premium horse hay is an art and a science, but there's only so much the farmer can do, because the success of a hay crop depends to a great extent on the weather. Try to buy the best quality hay you can afford, but realize that finding good hay during a "bad year for hay" can require diligent detective work. The best thing you can do to increase your chances of finding a year's supply is to learn to recognize top-notch hay when you see it.

Depending on the area and type of hay, a farmer will get from one to four crops, or cuttings, of hay per year. Horsemen are very opinioned on which cutting is the best to buy. Although there are some differences in the cuttings, the quality of the hay is much more important than the cutting. From a nutritional standpoint, all cuttings can result in prime horse hay.

Grass hays such as timothy, brome, and orchard grass cut at the optimum time have tight seed heads and an ideal nutrient content for horses. If the farmer waits until the plants are very tall and seed heads are mature, such as with some grass fields that are cut only once per season, the results are a high yield and a safe roughage, but a very low nutritive value. This is a good hay to feed to idle mature horses in the winter as part of their ration to keep them occupied without having to worry about them getting too fat.

With alfalfa, there will be some variation in protein content between cuttings. First-cut alfalfa hay has the reputation of having large, tough stems, but this is only true if the hay was too mature when cut. If first-cut hay is mowed at the prebloom stage, the stems will not be coarse and the nutritive value will be high. Weeds do tend to appear in first-cut hay.

Second-cut alfalfa hay is usually the fastest growing because it is developing during the hottest part of the season, and it usually has more stem in relation to leaf. Of all cuttings, second cut tends to be the lowest in crude protein, but its 16 percent average is adequate for most horses.

Third-cut (and later) alfalfa develops a higher leaf-to-stem ratio because of the lower growth during the cool part of the season. Therefore, third-cut hay will usually have the highest nutritive value. Horses that are not accustomed to a rich, leafy hay may experience flatulent (gaseous) colic or a loose stool.

Mixed hays, such as grass-alfalfa, are cut using the maturity of the alfalfa plants as a guide. For each day a plant stands after first flowering or past the boot stage, crude fiber increases and crude protein decreases by approximately 0.5 percent. The first cutting of a grass-alfalfa mix will contain a larger proportion of grasses than will subsequent cuttings.

Hay Quality

Good-quality hay usually has a bright green color and a fresh smell, but placing too much emphasis on color may be inappropriate and could cause you to turn down perfectly good hay. Although the bright green color indicates a high vitamin A (beta-carotene) content, some hays that are pale green or even tan on the outside due to bleaching may still be of good quality. Bleaching is caused by the interaction of dew or other moisture, the rays of the sun, and high ambient temperatures. Brown hay, however, indicates a loss of nutrients due to excess water or heat damage and should be avoided.

Be wary of hay containing poisonous plants, thistle, and plants with barbed awns such as foxtail or cheat grass, which can become embedded in the horse's mouth tissues.

Blister Beetles

Avoid hay containing blister beetles, because the beetles contain cantharidin, which is toxic to horses. Ingesting very small amounts of blister beetles can cause a horse severe pain and shock, and as little as 5 grams of beetles can kill a horse. Symptoms include depression, signs of abdominal pain, and frequent urination in small amounts. Confirmation of poisoning by blister beetles requires a laboratory analysis of the horse's urine or feces. There is no specific antidote for cantharidin, and the prognosis for recovery from poisoning is poor. Affected horses can be treated by a veterinarian with mineral oil and activated charcoal to minimize absorption of the toxin. Intravenous fluids can be given to correct dehydration and replace lost calcium and magnesium.

Blister beetles are found throughout the major alfalfa growing regions of the United States and are most prevalent during June and July. Therefore, in most areas the first cutting of alfalfa before June, and the last cutting from September on, will likely *not* have concentrations of beetles high enough to be a danger to horses. All hay containing blister beetles should be destroyed, because cantharidin is toxic to most other domestic animals, and storage of hay does not reduce the toxicity.

For more information on the incidence of blister beetles in your area, contact your county extension agent (see Appendix).

CAVEAT EMPTOR

You may get a good price on hay that's been rained on, but it could end up being a bad deal for your horse. If you have any doubt about the quality of the hay, cut open several bales to check inside before you buy.

10.1 LEAFY HAY

Good-quality hay should be leafy, fine stemmed, and adequately but not overly dry. Since two-thirds of the plant nutrients are in the leaves, the leaf-to-stem ratio should be high. The hay should not be brittle but instead soft to the touch, with little shattering of the leaves.

10.2 SOFT VS. DRY HAY

You should be able to roll the hay into a ball without much leaf loss. If you shake the hay, the leaves and flowers should stay on the stems and the hay should smell fresh, not moldy or caramelized.

"Sticks and powder" hay falls apart when you pick up a flake; all the leaves shatter and fall to the ground, and you're left holding the stems. Lost leaves mean lost nutrition. The best way to feed shattered hay is on a clean, hard surface such as rubber mats or concrete.

10.3 MOLDY HAY

Hay that is dusty, moldy, or musty smelling is not suitable for horses. Not only is it unpalatable, but it can contribute to respiratory diseases. Moldy hay contains fungal spores that generally appear as a whitish dust. They can be toxic to horses, causing colic or abortion. A person feeding moldy hay can develop allergic reactions to the mold and fungi and develop flulike symptoms. Some bales, like the one shown here, might be moldy clear through. Others bales might have mold in some of the flakes, while the rest of the bale is suitable for feeding. Your nose knows. If hay smells moldy, dusty, sour, or rancid, don't buy it, and don't feed it.

10.4 HAY TEMPERATURE

Bales containing too much moisture will mold, and can get so hot as to spontaneously combust, causing a barn fire. Check the internal temperature of a bale by cutting it open and passing your hand between some flakes. New hay will feel slightly warm as it cures. Avoid cured hay (more than 2 days old) that feels very warm, because even if the hay doesn't mold, heat can reduce the nutritional value of the hay.

10.5 HAY CUBES

Hay cubes, made from alfalfa hay that is chopped, processed, and compressed, are easy to store and feed. They are convenient to have on hand to feed to penned horses on windy days.

"CHOP"

Hay that is chopped, dried, and bagged, sometimes called "chop," makes a roughage of consistent quality that is easy to store and feed. Chop added to a grain ration can keep a horse from bolting his food.

Buying, Transporting, and Storing Hay

Purchasing enough hay to last your horses one year ensures you'll have hay when you need it. Whether this will work for you depends on the availability of hay in your area, how much room you have to store hay, and your budget. A year's supply of hay requires a large cash outlay, but it's usually cheaper in the long run to buy all your hay at harvest time, rather than buying smaller quantities throughout the year. Hay typically goes up in price as the winter wears on.

Once you find a good hay source, be it a local farmer or a hay dealer, establish a good long-term business relationship with him. Some farmers have large hay storage buildings, so if you don't have room to store a year's worth of hay at your barn, you might be able to store it for a few months at the hay grower's.

Special trucks called retrievers can pick up and transport stacks that contain approximately 5 tons (4.5 metric tons) of hay. Having hay delivered by the "stacker load" is very efficient and can be more cost-effective than hauling it yourself.

Hay sheds with an inside clearance of 18 feet (5.5 m) allow a retriever to back in and unload hay. Since the retriever sets the stack directly on the ground, you want to be sure that the site of the hay shed is higher than surrounding terrain to keep rain and snowmelt from running into the shed and spoiling the bottom bales.

10.6 FILLING A LOFT

Loading and hauling hay yourself can give you a feeling of accomplishment and save you the cost of delivery. Whether it's worth it in the long run depends on how many trips you have to make and how far you have to haul.

Getting hay onto your truck is only half the battle — the real fun is hoisting bales into the loft.

10.7 HAY ELEVATOR

An electric hay elevator can save a lot of wear and tear on your back.

BALE-MOVING TIP

It's sometimes easier to use a hay hook to drag a bale to where you want to stack it than it is to carry it.

10.8 BALE WAGON

Some farmers use bale wagons, or stackers, to pick up bales from the field and automatically place them in interlocking rows to form stacks. If you live fairly close to the farmer, he may use a stacker to deliver the hay right out of the field.

10.9 STACKING HAY

Once hay is delivered, it is often moved into a building or restacked outside on pallets. Bales typically weigh between 55 and 75 pounds (24.9 and 34 kg), so stacking them more than five high can give you a good workout.

10.10 INDOOR STORAGE

A large quantity of hay is best stored in a building separate from the horse barn for fire safety. Smaller quantities, up to a month's worth, can be brought to the barn and stored in a specially designed hay area or in a stall. This shows straw on the left and hay on the right.

10.11 OUTDOOR STORAGE

When storing hay outside, stack it on wooden pallets to prevent ground moisture from ruining the bottom bales. Feed stores and warehouses often give away used pallets. Cover the top and as much of the sides as possible to prevent spoilage from sun and precipitation.

10.12 WINTER PROTECTION

This 5-ton (4.5-metric-ton) load of hay was stacked on pallets and covered all the way to the ground with a heavy-duty tarp to protect the hay from drifting snow. Ropes over the top and around the stack prevent the tarp from billowing and ripping loose in strong winds. The ropes and at least part of the tarp have to be untied each time hay is needed from the stack. Check your tarp carefully — even the smallest hole in the tarp at the top of the stack can result in a funnel of moldy hay clear through to the bottom of the stack.

Heavy-duty tarps are expensive, but a lightweight bargain tarp seldom lasts even one season, especially if it's not tied down securely.

10.14 HAY SATCHEL

A hay satchel made from solid fabric such as nylon, canvas, or cotton, as shown here, is suited for dry, shattery hay because you won't lose valuable leaf in transit. The shoulder strap makes it easier to carry for a long distance when full.

10.13 FEED CART

A cart like this works great for delivering hay and grain rations at feeding time. With the short back gate removed and no buckets, a cart this size can hold two bales of hay.

HAY CARRIER

For carrying up to 30 pounds (13.6 kg) of hay to a distant pen or pasture, a simple hay carrier can be used. It's made from two 14-inch (35.6-cm) dowels and two pieces of ⅜-inch (9.7-mm) cord, 36 to 42 inches (91.4 to 106.7 cm) long. Measure 2 inches (5.1 cm) in from each end of the dowels and drill ⅜-inch (9.7-mm) holes. Pass a cord through each hole and tie a knot in the end of the cord.

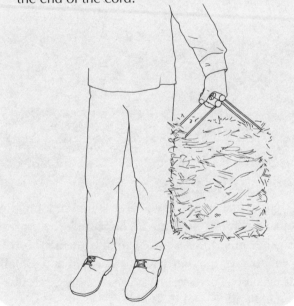

✦ GRAIN ✦

Grain should be used mainly as a supplement to hay, or roughage, which makes up the bulk of a horse's diet. Not all horses require grain. How much and what kinds of grain are fed depend on the quality of the hay, the horse's age, level of exercise, and condition, and whether the horse is breeding or lactating. One grain cannot cover all applications. There are grain mixes and pellets that are specifically formulated for horses of specific ages and performance levels.

Grains such as oats and corn can provide excess energy and cause some horses to become rambunctious and difficult to manage. Cutting back on the ration of such horses and allowing them more exercise will often calm them down. Always suit the grain ration to a horse's exercise requirements. Don't increase grain *before* increasing the amount of exercise; always *decrease* grain before decreasing exercise.

Consistent overfeeding is unkind and can easily make a horse overweight. This puts unnecessary stress on a horse's legs and feet, makes it harder for him to cool out, and makes him less nimble. If a horse eats too much grain at once (if he gets access to a grain barrel, for example) he could colic or founder, both of which can be life-threatening conditions. So make sure your grain is stored in such a manner that a horse can never get at it.

Bran is a by-product of milling wheat, rice, and other grains. A warm wheat-bran mash has traditionally been recommended for sick horses or those that have trouble eating because of dental problems. It also is sometimes used to add bulk and fiber to a horse' diet. Feeding wheat bran regularly can harm a horse because it is high in phosphorus, which binds with calcium, preventing absorption of calcium. This can lead to a dangerous calcium deficiency, which may result in bone deformities and lameness. A balanced diet can safely have five times as much calcium as phosphorus, but too much phosphorus can cause problems. Young horses in particular should rarely, if ever, be fed wheat bran. If a mash is needed, it can just as easily be made by soaking rolled oats, barley, or beet pulp in hot water.

Types of Grain

Whole oats are the mainstay of a horse's grain ration. They provide the right balance of fiber (from their hulls) and energy (from the kernels) to make them a relatively safe feed. Whole oats that are not chewed thoroughly can pass undigested through a horse, however, so oats are often "rolled" or "bruised" to crack their hulls and make them more digestible. Oats contain three times as much phosphorus as calcium, so they should be complemented by a calcium-rich hay.

Whole corn has a very thin covering that does not provide much fiber, and corn provides twice the energy of oats, so it can be too concentrated for some horses. Whole corn would be difficult for a young or old horse to chew, so the grain is often rolled or cracked into smaller pieces.

Rolled grains are easier to chew and digest but they also lose nutrients and can spoil more quickly than whole grains.

Barley is a high-energy grain, but because it tends to be dusty and has a prickly husk, it is usually mixed with other grains rather than fed straight.

Soybeans are an excellent source of high quality protein. Soy chips are beans that have been cracked.

Sweet feed mixes are usually made up of oats and/or barley, corn, and protein pellets. Molasses is added to bind the grains, cut down on dust, and make the feed more palatable. Sweet feed can spoil in hot climates, so buy fresh and store in a cool, dry place.

Pelleted grain rations are made of processed grains that have been ground and compressed.

Complete feeds are made from processed alfalfa hay, grains, and minerals. They come in wafers or "cakes" of various shapes and sizes. They provide a balanced diet so no additional hay or grain needs to be fed. Complete feeds are handy to feed on windy days when a horse's hay ration would likely blow away. It's best to feed long hay whenever you can, however, because complete feeds are eaten quickly and don't provide a horse with enough "chewing" time. A handful of wafers in a horse's normal ration can slow a horse down and keep him from bolting his feed (see photo 12.19). The quality of wafers varies and some crumble quite easily, resulting in a large amount of fines and powder in the bottom of the feed barrel.

11.1 GRAINS
Counter-clockwise from upper left: whole corn, rolled corn, rolled mix, whole oats, soybean chips (*center*).

11.2 PELLETED RATIONS
Clockwise from bottom left: sweet feed, a specialized pelleted grain ration (Equine Senior, Purina Mills, Inc., St. Louis, MO), and supplement (Farrier's Formula, Life Data Labs, Cherokee, AL).

11.3 COMPLETE FEEDS
Some wafers are solid (*upper left and bottom*), while others (*upper right*) crumble easily and turn into fines.

Mineral Supplements

Salt (sodium chloride), either loose or in block form, is the most common horse supplement.

▶ White blocks (plain salt) are fine for horses on good pasture or hay.

▶ Light red blocks have iodine added, for areas where soil lacks iodine.

▶ Dark red "trace mineral blocks" contain salt plus minerals such as iodine, zinc, manganese, iron, copper, and cobalt, which make up for poor roughage. These are fine for all horses. Minerals are added in small amounts so there's no danger a horse could "overdose."

▶ Trace mineral calcium/phosphorus (CA/P) blocks are fed to supplement and help balance the calcium and/or phosphorus in a horse's diet. If a horse's ration is heavy in phosphorus (grain), for example, he may need additional calcium. If you feed straight alfalfa hay and no grain, on the other hand, the horse's diet contains a lot of calcium, so you may need to provide more phosphorus. Trace mineral CA/P blocks are available in different CA/P ratios. Blocks containing 12 percent calcium and 12 percent phosphorus are called 12:12 or 1:1 blocks and are suitable for most horses. The safest practice is to provide each horse access to a plain salt block and a 12:12 block. Confer with your veterinarian or equine extension agent, who may suggest you have your hay and grain tested. This is especially important if you raise young horses.

▶ Brown blocks are usually molasses protein blocks that also contain salt. These "candy bar" blocks can disappear quickly and should be fed only if a horse's diet is low in protein. Eating too much at once could cause colic or diarrhea.

11.4 A horse should have free access to a trace mineral block. Do not place loose salt or a salt block in a horse's grain feeder. Not only will salt cause a metal feeder to rust very quickly, but a horse might eat too much salt or go off his feed.

11.5 A small wall-mounted mineral block holder is a handy way to provide supplement in a stall.

Psyllium and Sand Colic Prevention

Horses eating off the ground ingest dirt and sand. Sand can accumulate in a horse's intestines and lead to sand colic, a life-threatening intestinal obstruction. The best way to minimize ingestion of sand is to feed on clean rubber mats, grass, or snow. If you think your horse does ingest sand when he's eating, you can consider feeding psyllium, a high-fiber husk. Psyllium hydrophilic mucilioid contains 80 percent soluble fiber, which is thought to capture the sand and carry it through a horse's digestive system. Research is under way to measure the effectiveness of psyllium at removing sand.

Psyllium comes in powder, flakes, and pellets (see appendix) and is fed dry along with a horse's grain ration. If the horse is not getting grain as part of his regular diet, feed a small amount of grain, about a half-cup (125 mL), to carry the psyllium. Feed psyllium once a day for one week each month.

Grain Storage

Grain should be kept in a cool, dry place and fed as soon after it is purchased as is practical. Sweet feeds, because they contain molasses, can easily turn rancid. Cracked or rolled grains spoil more easily than whole grains, and all grains become moldy and rancid more quickly during hot, humid weather. During cold weather, feeds dressed with molasses can form hard clumps that are difficult to scoop.

Feeding moldy grain to your horse can cause heaves, liver damage, botulism, abortion, or death. Sometimes moldy grain has a whitish or black coating and will clump together. Moldy corn, however, can appear normal yet contain *Fusarium moniliform*, a fungus that produces toxins. These toxins cause severe neurological disorders and the highly fatal "moldy corn disease." As with hay, trust your nose, and discard any and all suspicious grain.

HOW TO DETERMINE IF YOUR HORSE IS INGESTING AND PASSING SAND

- ▸ Collect one fecal pile from rubber mats, grass, or clean bedding, not off dirt or sand.
- ▸ Place the sample in a 5-gallon (18.9-L) bucket and fill with water.
- ▸ Let soak, then stir until fecal material is completely broken apart.
- ▸ Dump water and "green stuff" off the top, leaving solid material at bottom.
- ▸ Keep rinsing and pouring off the water until all that's left at the bottom of the bucket is clean sand.
- ▸ If you collect ½ cup (125 mL) of sand or more, be glad your horse is passing the sand, but make immediate changes to your feeding practices to decrease his sand ingestion.

11.6 GRAIN HOPPER

Bulk grain can be delivered by a truck and stored in a hopper bin like this one. A cart is then used to haul grain for each feeding or to fill smaller storage containers in the barn.

Grain Storage

11.7 FEED ROOM

A dry, rodent-proof feed room is the ideal place to keep all grain, supplements, buckets, scales, and scoops. This makes regular feeding more efficient and results in minimal feed loss due to spoilage and vermin. The concrete floor in this room never gets damp, so extra bags of grain are stored directly on the floor until there is room in the barrels. If you don't have a separate room, plastic storage barrels with lids that fit tightly can keep grain clean and protect it from insects and rodents. To secure feed from marauding raccoons and gluttonous horses, install a positive locking latch on barrels and bins.

11.8 DIRT FLOOR

This barn has a dirt floor that gets wet during heavy rains and spring snowmelt. Sacks of feed are stored for a short period on pallets, until there is room in 40-gallon (154-L) plastic barrels. Because the barn has two resident cats, there is no problem with mice molesting the feed.

11.9 FEED BIN

A feed bin can be located in an alcove or at the end of an aisle. This bin is designed to store grain loose, in bags, or in barrels and to hold feeding scoops and buckets. A feed bin should be completely cleaned out periodically to remove grain and dust in the bottom corners that might mold or turn rancid. Slanted lids like these are good because they prevent items from being stored on top of the bin, which would make it difficult to open.

12 ♦ FEEDING PRACTICES ♦

It is part of the responsibility and satisfaction of horsekeeping to provide your horse with the proper amount and type of feed so he is fit and healthy. How much feed your horse needs depends on his size, condition, and activity level, and on the quality and type of feed you are using. A horse's digestive system is best suited for roughage like hay and pasture. Grass hay, such as timothy, brome, and orchard grass, makes excellent horse feed if baled and stored properly. Hay that is too mature, moldy, or dusty is not suitable for horses. Alfalfa is higher in protein, vitamins, and calcium than grass. It should be fed with care, because it can lead to mineral imbalance; too much protein might cause bone and kidney problems. Processed hay in the form of cubes is handy to feed and is especially useful during windy periods when loose hay would blow away. Many horses do not need to be fed grain. Horses that usually require grain include young horses, breeding stallions, pregnant mares, mares with foals, and horses doing hard work. Minerals can be supplemented in a horse's diet in the form of a trace mineral salt block or calcium/phosphorus block. Vitamins rarely have to be supplemented if you are feeding good quality hay. A horse should always have access to fresh water.

12 RULES OF HORSE FEEDING

1. Regularly measure the horse's weight with a weight tape.
2. Calculate how many pounds of hay he should be fed (see page 91).
3. Determine if he needs grain and calculate how many pounds of what type of grain is appropriate.
4. Don't overfeed. It's unkind and unhealthy.
5. Don't underfeed. A horse needs a certain amount of body fat.
6. Know what mineral supplements the horse needs.
7. Feed at least twice a day. Small amounts more often are better for the horse's digestive system.
8. Feed at the same times every day. Horses have a strong biological clock.
9. Be sure the horse always has fresh water.
10. Make feed changes gradually. This prevents gaseous and spasmodic colic.
11. Introduce a horse to pasture gradually to reduce the chance of colic and founder.
12. Never feed a horse that is hot from exercise. Wait until he cools down.

Calculating Hay Ration

How much hay a horse should be fed depends on the horse's weight, but it is also affected by the type and quality of the hay, the horse's age, size, and activity level, and environmental conditions. That's why the following equation uses a range to calculate the estimated hay ration according to a horse's weight. To know exactly how much you are feeding, feed by weight and not by volume.

Weight of horse x 1.5–1.75% = weight of
   daily hay ration
Daily hay ration/Number of feedings per
   day = lb (kg) per feeding

Example:

1,000 lb (454 kg) x .016 = 16 lb (7.3 kg)
16 lb/2 = 8 lb (3.6 kg) per feeding

A standard tape measure and the following chart can also be used to estimate a horse's weight.

HEART GIRTH VS. WEIGHT

| HEART GIRTH | | WEIGHT | |
in	cm	lb	kg
40	102	200	91
45	114	275	125
50	127	375	170
55	138	500	227
60	152	650	295
62	157	720	327
64	163	790	358
66	168	860	390
68	173	930	422
70	178	1000	454
72	183	1070	485
74	188	1140	517
76	193	1210	549
78	198	1290	585
80	203	1370	621

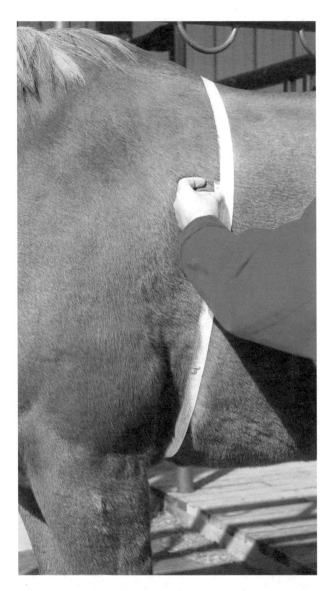

12.1 There is a consistent correlation between a horse's heart girth and his weight. A specially marked "weight tape" enables you to get a fairly accurate estimate of a horse's weight by measuring around the horse's barrel just behind the withers and front legs.

Weighing Hay and Grain

The flakes (slabs, leaves, slices) that make up a bale of hay can vary in weight from 2 to 7 pounds (0.9 to 3.2 kg) depending on the type of hay, moisture content, thickness of the flake, and how tightly the hay was baled. A hanging scale works well for weighing hay portions at each feeding (photo 12.2). After several feedings, you'll develop a feel for the weight of the flakes of hay you're feeding, but it's still a good idea to check a ration on the scale periodically, to prevent feeding too much or too little. When the temperature drops, be sure to increase your horse's hay ration (see cold weather worksheet, page 94).

The area where you feed hay from should be well lighted, so you can check each flake carefully as you remove it from the bale. Look for wire, plastic, and sticks that might have been picked up by the baler, and be especially vigilant for mold and the carcasses of blister beetles (see chapter 10).

As with hay, grain and concentrates should be fed by weight, not by volume (photo 12.3). Before relying on a scoop or can to measure rations, know the weight of each feed it holds. To find out, weigh the empty scoop or can, then fill it with the grain or concentrate you are feeding and weigh it again. Subtract the empty weight from the full weight to find the weight of the grain. Do this for each type of grain or concentrate you feed, as the weight will vary widely between rolled oats, a pelleted feed, and whole corn, for example. Draw marks on a can, or use a calibrated scoop to help you more accurately measure grain by weight.

Evaluating a Horse's Condition

The "condition" of a horse can refer to his fitness, but here it refers to how much fat he's carrying. A horse needs a certain amount of fat for optimum health and performance, but too much fat can stress a horse's legs and feet, put extra load on the heart and lungs, make it difficult for a horse to cool out, and make a horse heavy on his feet. Palpate, or feel, the back, ribs, neck, shoulder, withers, and tail head. Most horses should be maintained at a condition score of between 5 and 6 (see chart on page 93).

12.2 Use a hanging scale to weigh hay.

12.3 Grain should be fed by weight, not volume.

SCORE	DESCRIPTION
1 Poor	Extremely emaciated; no fatty tissue can be felt; ribs, vertebrae, and withers projecting prominently; bone structure easily noticeable
2 Very thin	Emaciated; ribs and vertebrae prominent; faintly noticeable bone structure; very thin neck, shoulders, and withers
3 Thin	Ribs can easily be seen; slight fat cover over ribs, neck, shoulders, and withers yet they appear thin
4 Moderately thin	Very faint outline of ribs noticeable; neck, shoulder, and withers are not obviously thin; negative crease (prominent vertebrae) along back
5 Moderate	Ribs cannot be seen but can be easily felt; fat around tailhead beginning to feel spongy; back is level over loin; shoulder blends smoothly into body; withers rounded
6 Fleshy	Can barely feel ribs; fleshy tailhead feels spongy; may be slight crease down back over loin
7 Fat	Individual ribs can be felt, but noticeable filling between ribs with fat; crease down back over loin; fat deposited along neck, withers, tailhead, and area behind shoulders
8 Very fat	Fat deposited along inner buttocks; difficult to feel ribs; thickening of neck; fat withers, tailhead, and area behind shoulders; positive crease down back over loin
9 Obese	Can't feel ribs; bulging fat on neck, withers, tailhead, area behind shoulders, and along inner buttocks, which may rub together; very obvious crease down back over loin; flank filled in flush

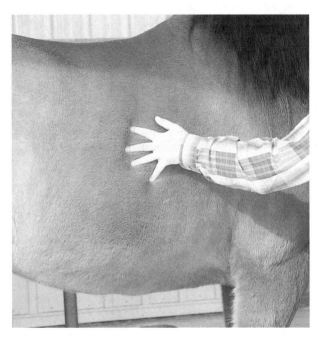

12.4 FEEL THE RIBS

A long hair coat will interfere with visual observations, so it's important to evaluate the horse by feel as well. As you palpate the rib cage, be careful not to mistake fat for the fit muscle tone of a well-conditioned horse.

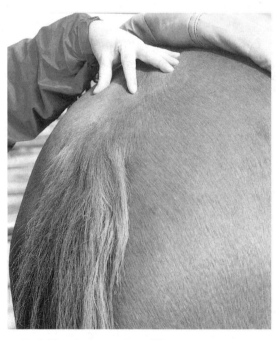

12.5 PALPATE THE TAILHEAD

The soft area above the tail is one of the best places to check a horse's level of fat. Look for the presence of a crease down the horse's back.

Guideline for Cold-Weather Hay Increase

When the temperature drops it's better to increase the hay ration than to feed more grain. Grain is digested relatively quickly and the heat from its digestion is soon gone. Hay generates heat over a longer period of time, helping to keep the horse warm longer. *Rule:* Increase your horse's hay ration by 10 percent for every 10 degrees Fahrenheit (6°C) below freezing (32°F or 0°C).

12.6 Labeling buckets with the horses' names minimizes confusion and helps ensure each horse gets his correct ration.

12.7 This feed room has a shelf along the back wall that's at a convenient height for weighing and measuring grain. Plastic barrels for grain are set on 12-inch (30.5-cm) platforms, for convenience when scooping grain. Individual rations are calculated and written on an erasable white message board, which is easy to update. This helps keep rations current, especially for persons you ask to do chores.

CALCULATING COLD-WEATHER HAY INCREASE

DATA	EXAMPLE 1	EXAMPLE 2		
Current temperature	22°F (−6°C)	−12°F (−24°C)		
Degrees below freezing	10°F (−12°C)	40°F (4°C)		
Percentage to be increased	10%	40%		
Normal hay ration	16 lb (7.3 kg)	16 lb (7.3 kg)		
Additional hay	1.6 lb (0.7 kg) (16 x 0.10) (7.3 x 0.10)	6.4 lb (2.9 kg) (16 x 0.40) (7.3 x 0.40)		
Adjusted cold-weather ration	17.6 lb (8 kg)	22.4 lb (10.2 kg)		

Feeding Areas

A horse's anatomy is designed to ingest food at ground level, so it's best to feed horses on the ground if possible. The feeding area should be free of manure to minimize parasite infestation. In areas with dirt or gravel footing, a clean place to feed, such as on rubber mats, cuts down on wasted feed and minimizes ingestion of sand and gravel that could lead to colic (see Psyllium and Sand Colic Prevention, page 88). On windy days, a sheltered feeding area should be provided to keep hay from blowing away.

12.8 FEEDING ON THE GROUND

Feeding on the ground is the most natural and best alternative, especially if there is some grass cover, as shown here on a dry pasture. Feed the horse in a different area before dirt starts showing through the native grass cover.

12.9 FEEDING ON THE SNOW

In winter, pastured horses can be fed right on the snow or frozen ground. If the snow is more than an inch (2.5 cm) deep, clear an area to bare ground with your boot to keep the hay leaves from getting lost and to shelter the hay from wind.

When feeding two or more horses, the feeding area should be large enough to put out more piles of feed than there are horses. That way, even if the top horses in the social hierarchy dominate more than their share of feed piles, the lower horses will still have access to feed.

Feeding Areas (continued)

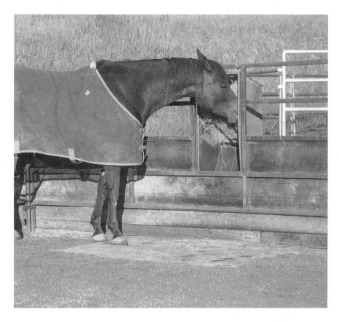

12.10 Rubber Mat and Wind Screen

A horse can ingest an incredible amount of sand and dirt if fed on loose soil. This foreign matter can accumulate in his digestive tract and cause painful, life-threatening, and expensive bouts of sand colic (see Psyllium and Sand Colic Prevention, page 88). A 4-foot by 6-foot (1.2-m by 1.8-m) rubber mat placed over the dirt or gravel under a hay feeder catches hay leaves that fall from the feeder. The back of this feeder faces west, which is the direction of the prevailing wind. To prevent hay from blowing away, a 36-inch-wide (0.9-m) section of ¼-inch (6.4-mm) rubber conveyor belting is wired to the middle rail of the pen with 11-gauge galvanized wire. To hold the belting in place and seal the bottom edge of the windscreen, railroad ties are placed at the base. This custom hay and grain feeder is mounted on the *outside* of the pen, where the horse can't rub on its edges.

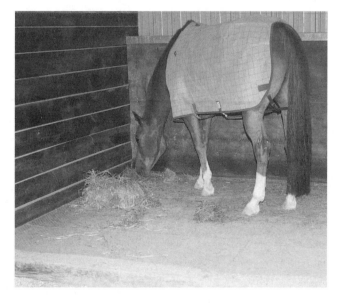

12.11 Feed Mats

Here, rubber mats provide a smooth, clean surface on which to feed hay at ground level. Grain is fed in a dish that is removed after each feeding. The mats cover a 10-foot by 10-foot (3-m by 3-m) area under a roof overhang on the sunny south side of the barn. The feeding area is separated from the gravel portion of this pen by railroad ties. Each tie is staked to the ground with two ⅝-inch (15.7-mm) by 16-inch (40.6-cm) steel rods. The mats need to be swept about once a week to remove gravel tracked onto the area by the horse's hooves. In stalls, rubber mats can be swept clean of bedding in one corner for feeding hay on the floor.

Making Mat Anchors

Here's a simple way to make horseproof anchors to help keep even a four-corner junction in place. The rounded stake head will seat tightly against the mats and allow a broom, shovel, or hoof to slide over it without catching the edge.

Materials

▸ ¼-inch (6.4-mm) diameter steel rod approximately 12 to 18 inches (30.5 to 45.7 cm) long (the softer the soil, the longer the rod)

▸ ¼-inch (6.4-mm) steel washer

▸ 1½-inch (3.8-cm) diameter fender washer with a ⁵/₁₆-inch (7.9-mm) center hole. (Fender washers are relatively thin and have a large outer diameter compared to a small center hole; available at hardware and auto stores.)

▸ 1-inch-diameter (2.5-cm) pipe (See step 2.)

Instructions

1. Weld the ¼-inch (6.4-mm) washer flush with the end of the rod and then weld the fender washer on top of the smaller washer. Weld the center hole of the fender washer and two or three spots around the smaller washer on the underside.

2. To round the head, insert the stake through a piece of 1-inch (2.5-cm) diameter pipe so the head of the stake sits flat on the pipe. Bend the perimeter of the large washer down over the edge of the pipe using light taps with a hammer. (*Always* wear eye protection whenever striking metal with a hammer.) Remove the stake from the pipe.

3. Clean out debris from under the corners of the mats so they lie flat and even. Insert the mat stake between the corners of the mats and use a hammer to drive it to the surface of the mats. Keep the stake vertical so the head will seat level.

Feeders

Most horses prefer grain to hay, so if hay and grain are fed in the same container, they will pull the hay out onto the ground to get at the grain. It's best to feed hay and grain separately and to keep both feeds well away from the water supply to prevent the horse from fouling the water with feed.

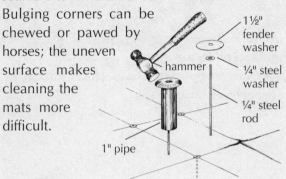

MAT ANCHORS

Mats, especially those thinner than ¾ inch (19.1 mm), have a tendency to bulge up along seams and where corners intersect. Bulging corners can be chewed or pawed by horses; the uneven surface makes cleaning the mats more difficult.

1½" fender washer
hammer
¼" steel washer
¼" steel rod
1" pipe

Hay/grain feeders mounted at head level might keep most of the hay from falling to the ground and being trampled, but hay stems and seeds can irritate the horse's eyes as he dives deep for tasty morsels. Also, the forelock and face get covered with hay flakes, requiring extra grooming. Some feeders have nooks and crannies that trap bits of feed. This might cause a hungry, determined horse to paw the wall or pen, or bite at the feeder in frustration, both of which can turn into habits. If the trapped feed is not removed daily, it can mold and, if moist, can cause the feeder to rust.

12.12 AVOID WOOD FEEDERS

Horses will almost certainly chew feeders made of wood, so metal or durable plastic feeders are preferable. This feeder poses a hazard to the horse's gums and digestive system from wood splinters and also invites injury from exposed nails.

Feeders

12.13 GROUP HAY FEEDER

When feeding a group of horses, a round metal feeder like this one is a good way to prevent a bully from monopolizing all the hay.

12.14 FEED JUST ENOUGH

Although heaping a hay rack might mean you won't have to feed the horses for a week or more, a large percentage of the hay will likely get rained on, pulled out, or trampled into the dirt. For less waste, give horses only what they will clean up in one feeding without leaving a trace. Feeding twice or more each day also gives you an opportunity to check the horses over. A serious problem that's not noticed for a few days can mean the loss of a horse.

12.15 SWING-OUT FEEDER

With a swing-out hay and grain feeder, you don't have to enter the stall to feed. This feeder hinges at the corner of the stall and the entire unit swings out, which allows you to put hay and grain in the feeder without entering the stall. Once filled, the feeder swings back into the stall, and is locked in place by a latch on the outside of the feeder door.

12.16 CORNER HAY FEEDER

This plastic corner feeder is mounted low, so the horse eats in a more natural position. The feeder is 12 inches (30.5 cm) off the floor to allow for easy sweeping. There is a removable plug in the bottom for washing and draining.

12.17 CORNER GRAIN FEEDER

Another stall-feeding system uses a plastic corner grain feeder. Hay is fed on the floor. There is a small feed door in the grille that allows access to the grain dish from outside the stall; hay still must be carried in.

12.18 NO-SPILL GRAIN BUCKET

Some horses root around in a grain dish and push half the grain out onto the floor. This no-spill corner bucket has a removable plastic rim that overhangs the inside of the bucket to prevent grain from being pushed out.

12.19 THE IDEAL GRAIN DISH

A. If you feed a moderate to large amount of grain in a small, deep bucket, it is easy for the horse to bolt (gulp without chewing) his feed or push it out of the bucket onto the floor or ground.

B. This large, soft rubber dish spreads the feed over a larger area,

which will slow the horse down. A few rocks in the grain will cause the horse to search and browse the grain, rather than gobble. Horses often treat this type of grain dish as a toy, however, something to pick up, throw, chew, and generally abuse.

C. This large, hard plastic feed dish has tapered sides that make the dish difficult for a horse to pick up or tip over. It is ideal for pen and pasture feeding and lasts a long time. Wafers added to the grain will encourage the horse to sort, sift, and chew his feed thoroughly.

Feeders (continued)

12.20 TIRE FEEDER

This hay and grain feeder consists of three truck tires fastened together. The bead, or inner rim, of the top tire has been cut out to allow a horse to reach grain placed inside the tire. Hay is dropped in the center. Even though tire feeders are virtually indestructible, they are difficult to keep sanitary. Be alert for signs of chewing, because ingested rubber could cause colic.

12.21 HAY BAGS

Hay bags are a useful alternative to feeding on the ground during wet or windy weather when the feeding area has no shelter. A hay bag must be made of extra-tough material and sturdy construction because horses are vigorous eaters.

12.22 HANGING A HAY BAG

A hay bag should be secured high enough so a horse can't get a foot caught in it and should be fastened to the pen at the bottom as well as the top. The bag shown here is not secured at the bottom so the horse could easily flip the bag over the top rail to the outside of the pen and then be unable to reach it.

12.23 HAY NET

Hay fed in a net, or a double net, will occupy a horse longer than hay fed on the ground or in an open manger. In some cases, using a hay net in sheltered areas is better than feeding on the ground, since it minimizes the amount of hay a horse pushes out into the wind. Make sure the net is well secured and high enough so a horse can't get a foot caught in it. A hay net is handy for soaking hay that is a bit dusty (i.e., slight field dust, but *not* dusty from mold; *never* feed moldy hay).

12.24 NET WITHIN A FEEDER

In outdoor situations during wet or windy weather, putting hay in a net *within* a feeder prevents a horse from pulling out huge sections of hay that can be blown away or trampled into the mud. For bored horses in a stall, double hay nets within a feeder will keep the horse occupied and help minimize vices caused by boredom. Make sure the hay net is tied to the feeder to prevent a horse from pulling it out and getting tangled.

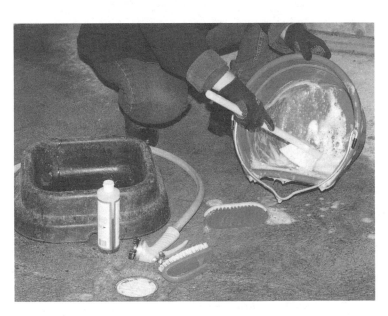

12.25 KEEP FEEDERS CLEAN

Feed dishes and water buckets should be scrubbed with hot soapy water periodically. Be certain to rinse off every bit of soapy residue and let the buckets and feeders dry in the sun.

✦ WATER ✦

One of the most important responsibilities of a stable manager is to ensure horses are provided with an adequate supply of fresh drinking water. A horse drinks between 5 and 20 gallons (18.9 and 75.7 L) of water per day, usually during two drinking sessions. The most typical time for a horse to drink is after he's eaten the majority of his hay ration, so water should be available at those times, but it's a good idea to have water available 24 hours a day. A horse will drink more water during hot weather than during cold weather, and horses that are lactating or exercising will require more water. Also, eating alfalfa hay and salt will cause a horse to drink more water. If a horse "goes off his water" and doesn't drink, or is deprived of water, he can suffer impaction colic (a blockage of the intestines by dry feed) or dehydration, both of which can be life-threatening.

Water doesn't have to be warm. The ideal is 50°F (10°C), which is about the temperature of water coming through underground pipes. Water should be fresh and safe, whether from a natural source, such as a creek or a well, or from a municipal water supply. You should periodically have the water tested for contamination.

If you use a waterer, whether a tank, tub, or bucket, it should be at least several steps away from hay and grain feeders to minimize pollution of the water with feed. If outdoors, the waterer should be sheltered from prevailing winds to keep wind-blown dust and debris from fouling the water. Waterers should not be located in fenced corners where a drinking horse might be trapped by other

horses. A waterer should be rustproof and easy to dump and scrub clean, and should be cleaned frequently to remove dirt, algae, slobber, and particles of hay and grain. A plastic bristled toilet brush is an excellent, inexpensive tool for scrubbing waterers because the curved shape enables you to scour the bottom contours.

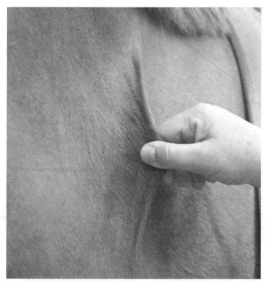

13.1 To check if a horse is dehydrated, use the pinch test. Pick up a fold of skin in the neck/shoulder area and pull it away from the horse's body. When you release the fold of skin, it should return almost immediately to its normal flat position. If the skin remains markedly peaked for 2 to 3 seconds, it likely indicates a degree of dehydration. A "standing tent" of skin that lasts 5 seconds or more indicates moderate to severe dehydration that might require the attention of a veterinarian.

Natural Sources

13.2 PONDS AND CREEKS

Water standing in pastures or draining into creeks and ponds can contain toxins from fertilizer runoff. Have water tested before allowing horses free access to it.

Keep an eye on ponds for algae buildup, because when algae die they produce a toxin that can be poisonous to horses. Horses that drink consistently from ponds and creeks with sandy bottoms can ingest enough sand to cause colic. Dig out the drinking spot and line the bottom with rocks. Check ponds and creeks during subfreezing weather to make sure there is an open place for horses to drink (see chapter 16). To avoid polluting a pond or creek with manure, you can fence off the water and set up a gravity feed pipe to fill a tank from the water source.

13.3 DON'T RELY ON SNOW!

Don't expect your horse to get enough water from snow. If snow is 2 inches (5.1 cm) deep, a horse would have to completely vacuum up an area 1 foot (0.3 m) wide by 56 feet (17.1 m) long to get 8 gallons (30.3 L) of water! Plus, he'd have to use precious body heat to melt all that snow.

Automatic Waterers

A properly installed automatic waterer is a convenient means of providing unlimited fresh water for horses. If not installed properly and monitored regularly, however, an automatic waterer can give you a false sense of security. Unless you actually see your horse drinking, you don't know if he is using the waterer, and even then you can't monitor how much water he is drinking unless the waterer is equipped with a meter. You might assume your horse is getting sufficient water when, in fact, he is not. Or you might go to the barn one day and find your horse's stall has become a wading pool because the waterer shutoff valve has malfunctioned or a water pipe has frozen and burst. Be sure to check the units twice a day to make sure they're functioning properly. A failed heater or a stuck valve can cause very expensive, time-consuming problems as well as thirsty, dehydrated horses. Be sure to install an accessible water shutoff valve somewhere between the main waterline and the waterer. If anything goes wrong with the waterer, you want to be able to quickly turn off the water and leave it off until the repair is completed.

Automatic Waterers

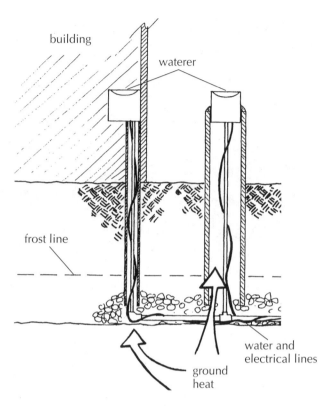

building

waterer

frost line

water and electrical lines

ground heat

13.4 PASTURE WATERER

Round waterers that mount on the end of a pipe set into the ground are especially safe for horses. They are usually mounted on a plastic pipe, as seen here, or on a concrete pipe. Horses prefer waterers that fill automatically as they drink, rather than cattle-type waterers that are activated by a paddle.

Even if an automatic waterer is in an enclosed building, the supply pipe should be buried below the frost line. The pipe is enclosed in a larger, empty pipe that draws warm air from the earth below the frost line. So even without electric heat, water can remain ice-free in temperatures around 20°F (−6.6°C). In colder climates, a heat cable or built-in heater can ensure liquid water at any temperature.

MOUNTING A STALL WATERER

Some automatic waterers are designed to mount on a stall wall. The wall must be thick enough and solid enough to securely hold the fasteners attaching the waterer to the wall.

Tanks and Barrels

One of the most common methods of watering horses is with tanks and barrels. The water should always be fresh and clear. Algae buildup can be dealt with by frequent emptying and cleaning, by adding carp to the tank to feed on the algae, or by adding an algaecide to the water (see Appendix).

13.5 STOCK TANK
This 100-gallon (378.5-L) galvanized steel stock tank has been in active use for more than 20 years. With only one horse using it, a tank this size only has to be scrubbed and filled about once a week.

13.6 HALF BARREL
Lightweight, durable water tubs can be made from 50-gallon (189.3-L) plastic barrels. Use a hand saw, circular saw, or reciprocating saw to cut a barrel in half for two 25-gallon (94.6-L) tubs. These tubs are very easy to clean, and ice up to ½-inch (12.7-mm) thick can be broken by kicking the side of the barrel. Here, large rocks keep a horse at a distance so he can't defecate in the tub, paw the water, or knock the tub over.

13.7 WATER BARREL
A plastic barrel can be cut taller for more capacity and to keep horses from pawing in the water, as some horses tend to do with shorter tubs. A 10-inch-wide (25.4-cm) top section of the barrel can be left when cutting. Drill two holes in this tab so a length of rope can fasten the tub to the pen. This prevents the barrel from being knocked over when it's almost empty. A panic snap can connect the rope for easy release and attachment when dumping and cleaning the tub.

Water Buckets

Buckets hung on the walls are a common way to provide water in stalls. Setting a bucket on the floor is not a good idea because it will likely be knocked over. The hanger should hold the bucket securely to prevent water from splashing out. The bucket should be easy to attach and remove from the hanger, even when wearing winter gloves. Buckets should be removed and rinsed out with every water change, and scrubbed when residue can be seen on the inside.

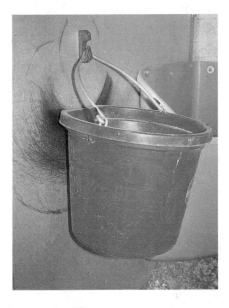

13.8 BUCKET HOOK
The bail of this rigid plastic bucket snaps into a plastic holder that's screwed to the wall. The handle is held securely, but you can see by the marks on the wall how much this single point of attachment allows the bucket to swing from side to side.

13.9 BAD SYSTEM
A drawback to buckets is that some horses don't respect them. This one was squashed by a rubbing horse, and now the handle is uncomfortable to hold. The three jury-rigged snap-and-twine attachments are cumbersome to operate and the twine encourages a horse to chew.

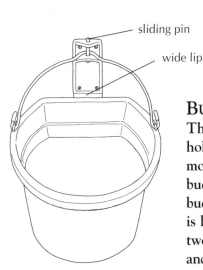

sliding pin

wide lip

BUCKET HOLDER
This cast aluminum flat-wall bucket holder, which also comes in a corner-mount style, holds a plastic or rubber bucket by two points. The rim of the bucket hooks on a wide lip, and the bail is locked in place by a sliding pin. This two-point attachment is easy to operate and holds the bucket very securely.

TIP
If a horse develops the unpleasant habit of defecating in his water bucket, try placing large rocks on the floor beneath the bucket to keep the horse from backing up to it.

Winter Tanks and Barrels

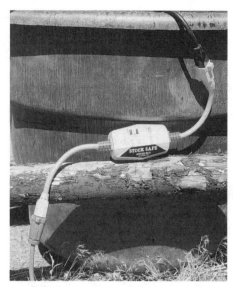

13.10 GFCI

Plugging a tank heater into a ground fault circuit interrupter (GFCI) extension cord or receptacle can prevent shocks. A GFCI interrupts the circuit, or shuts down, when a ground fault (current leakage) occurs. Unlike a fuse or circuit breaker, a GFCI is triggered by miniscule amounts of electricity, 5 milliamps, and shuts down in as little as 1/40 of a second. Check the heater daily, because if a short in the heater or wiring has tripped the GFCI, the water could freeze.

13.11 DANGER!

A tank located near a building or post must be placed so it can't trap a horse's leg. The tank should be either right against the structure or at least 16 inches (40.6 cm) away. It's dangerous to have a hydrant in a horse pen because a horse could turn it on or get injured on it. Be sure to break the ice in the winter so your horse can drink (see photo 16.11).

TANK HEATER

Electric tank heaters either sit on the bottom of a water tank or float on the surface of the water. Most have a thermostat that ensures the water doesn't get too hot. Many can't be used with extension cords, which limits their use. It's important but difficult to make sure a horse cannot get at the heater or the cord.

If the tank is large enough, half of the top can be covered by plywood and the heater installed under the cover, as illustrated here. The electric cord powering the heater can be run through a steel or thick-walled plastic pipe that's fastened to the rails of the pen. When using a tank heater, keep a close eye on your horses' drinking routines and touch the water with your hand frequently. If the water has a charge, it will punish your horse when he tries to drink.

Winter Water Buckets

13.12 ELECTRIC FREEZEPROOF BUCKET

Freezeproof buckets use thermostatically controlled wires embedded in the walls of the bucket to keep water between 40°F (4.4°C) and 60°F (15.5°C). The cords on heated buckets are only 4 feet to 6 feet (1.2 to 1.8 m) long, so an outlet may have to be installed, or a heavy-duty extension cord used, to allow the bucket to be located where it's needed. The cord should exit the stall wall through a hole as near the bucket as possible to prevent a horse from chewing on it. The cord is permanently attached to the bucket, and must be unplugged from the outlet and pulled through the wall to remove the bucket for cleaning. Some need a special bucket holder, while others, like the model shown here, attach to a standard two-point bracket, as shown on page 106.

13.13 INSULATED BUCKET HOLDER

An insulated bucket holder uses no electricity and some hold a standard 5-gallon (18.9-L) bucket. Insulation in the walls and bottom of the holder conserves heat from the water to delay freezing. The colder the air, the more frequently the water in the bucket needs to be changed to keep it from freezing. Unless it is extremely cold, an insulated bucket holder will keep water ice-free for at least several hours.

BUCKET HEATER

A drop-in electric heater can be used to quickly warm a bucket of water before setting in an insulated holder, or to heat a small quantity of water for washing a horse, treating a wound, or thawing a frozen hose. There are several types of drop-in bucket heaters, available in most tack catalogs. They should *not* be used to keep a horse's water from freezing.

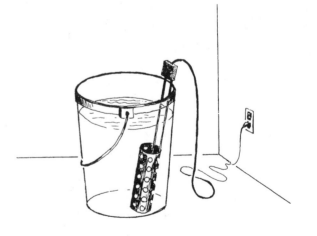

Hydrants and Faucets

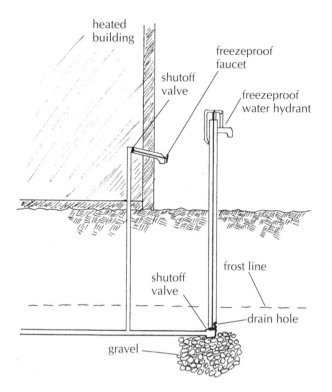

heated building

freezeproof faucet

shutoff valve

freezeproof water hydrant

shutoff valve

frost line

drain hole

gravel

FREEZEPROOF HYDRANTS AND FAUCETS

It's a good idea to install freezeproof water hydrants, even if you live in an area that seldom freezes. The shutoff valve is located below the frost line at the bottom end of the vertical water pipe or stem. When the hydrant is turned off, water in the stem drains out a hole next to the valve into a pocket of gravel. This prevents water from freezing in the stem and bursting it. Freezeproof hydrants are available in a variety of lengths to accommodate frost depths in different areas.

A freezeproof faucet works on the same principle as a freezeproof hydrant, by draining water from the section of pipe at risk of freezing. Instead of the shutoff valve being located near the handle, it's at the end of a pipe stem that projects into a heated building. The stem slopes slightly downward toward the spout of the faucet, so when the water is turned off all the water drains from the stem. If the building isn't heated, even a freezeproof faucet is likely to freeze.

COLD-WEATHER TIPS

▶ A simple but often overlooked precaution to help keep hydrants and faucets from freezing is to always disconnect the hose when you're through with it, so water can drain from the mouth of the hydrant or faucet.

▶ To prevent a hose from freezing, drain it thoroughly after each use. Pass one end of the hose over a tall rail, fence, roof joist, or other high object, and pull the entire length of the hose slowly toward you so all the water runs out the other end. The higher and the slower you go, the better chances are that the hose will be ice-free the next time you use it.

▶ To thaw out a hose that's frozen, coil it in an empty water tub and cover it with hot water. If you don't have hot tap water in your barn, use cold water and then warm it with a drop-in bucket heater (see page 108).

Mud Prevention

A waterer should be higher than the surrounding ground to encourage rain and snowmelt to drain away and to minimize mud around the base of the waterer. If practical, move the tank or tub periodically to let the land recover.

HYDRANT FLAG

An eye-catching flag on the hydrant handle will tell you at a distant glance when the hydrant is running — and whether you have remembered to turn it off. Affix the stick extension (about 36 inches [91.4 cm] long) to the handle of the hydrant with tape or large rubber bands. Tape a piece of weather-resistant bright fabric, such as plastic "flagging" ribbon used by surveyors or a red plastic flag that is free from some building supply stores, to the end of the stick.

13.14 MUD DESTROYS HOOVES

The area around waterers gets a lot of traffic, which can destroy the vegetation and turn the area to mud. Forcing a horse to stand in mud by purposely overflowing a water trough is a *bad* idea, as it leads to deterioration of hooves (see Mud Control, chapter 7).

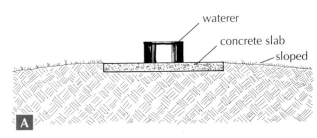

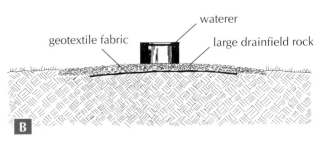

WATERER MUD CONTROL

A 12-foot by 12-foot (3.7-m by 3.7-m) concrete slab (A) is a sure way to prevent mud. The concrete should be at least 4 inches (10.2 cm) thick and textured for traction (see photo 1.8). This works best for automatic waterers that are attached to the concrete or for large, heavy tanks that horses can't slide off the slab.

An alternative is to cover the area with large, round drainfield rock, 2½ inches (6.4 cm) in diameter or larger (B). Large rocks will be less likely to disappear into the mud, and will be less likely to pack in horses' hooves and be carried from the area during wet weather. To prevent a stone or gravel layer from mixing with the soil below, first put a down a geotextile fabric layer. This permeable polypropylene fabric lets water drain through and prevents fine dirt from migrating upward into the gravel. For availability, contact local excavators.

Hoses

Having essential hose repair items on hand will minimize inconvenience and save the cost of a new hose if one springs a leak or gets chewed by a horse, or if the end gets flattened by a vehicle. The most common hose diameters, measured inside the hose, are ⅝ inch (15.7 mm) and ¾ inch (19.1 mm). Quick connectors, valves, and nozzles make hoses easier to use and conserve water.

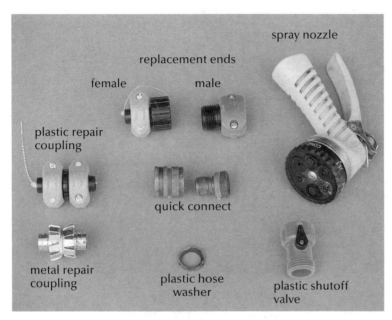

spray nozzle

replacement ends

female   male

plastic repair coupling

quick connect

metal repair coupling

plastic hose washer

plastic shutoff valve

13.15 HOSE REPAIR
On the left are two types of couplings for making permanent hose repairs: a metal one with tabs that crimp onto the hose, and a plastic one that clamps on with screws. At the top are replacement ends, female on the left and male on the right. In the middle is a brass "quick connect"; one piece screws onto a faucet, the other onto a hose. At bottom center is a plastic hose washer; a bad washer is often the cause of leaky hose connections. At top right is an adjustable spray nozzle, with six settings. At bottom right is an in-line plastic shutoff valve; no hose should be without one.

REPAIRING A HOSE

A quick, clean way to cut the damaged section out of a hose is to set the hose on a block of wood and cut it with one swift blow of a hammer and wide wood chisel. A sharp knife, tubing cutter, or heavy shears will also work.

To help the plastic repair attachment slide into the hose more readily, use dish soap, shampoo, or saliva. Once the piece is started, tap the end with a block of wood to drive it home.

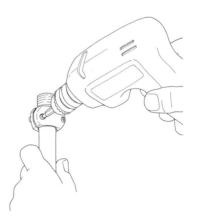

A cordless screwdriver or drill makes quick work of tightening the screws in the collars of hose repair pieces.

Hoses (continued)

13.16 Hose Rack

A hose lying in a heap on the ground takes up a lot of space and can easily trip a person or horse. It is also more likely to be damaged by traffic and tools. Hoses neatly coiled and hung on the wall or wound onto a specially designed spool are easier to use and will last longer.

13.17 Hose Hanger

A handy hose hanger for outdoor situations is the "clothespin post." A wooden support post set next to a hydrant is a good idea in any case, to protect the hydrant from being bent and a horse or person from being injured in the event of a collision. The top of this support post, which is 5 feet (1.5 m) from the ground, was carved with a chain saw to form a cradle in which to hang the hose.

Splash Cup

A "splash cup" will prevent water from blasting on the ground and splashing mud on your boots and pants.

    1. Cut a piece of 3-inch-diameter (7.6-cm) or larger PVC pipe to a length of 12 inches (30.5 cm).

    2. Bury the pipe vertically in the ground, centered on the exact spot where water from the hydrant hits the ground.

    3. Extend the pipe about 1 inch (2.5 cm) above the ground to help keep dirt out.

    4. Fill the pipe with gravel or rocks that are large enough not to be washed out by the force of the water.

splash cup

✦ SAFETY ✦

Safe facilities and management practices will help prevent injuries to you and your horse, minimize damage to the facilities, and enhance the security of your operation.

Handling Horses

Safe horse handling is covered in depth in other books (see Appendix), but here are a few reminders that bear repeating:

- ▶ Wear safe clothing: shoes with substance, not sneakers, and definitely not sandals.
- ▶ Avoid loose clothing that could blow around.
- ▶ Keep long hair tied back so it doesn't obstruct your vision and won't get caught.
- ▶ Pay attention to the horse and to the surroundings for things that might spook a horse and cause him to jump into you.
- ▶ When leading through stalls and gates, be sure the doors and latches are fully opened to prevent the horse from catching a blanket or getting cut or wedged in the doorway.
- ▶ Never turn out a horse that's wearing a halter.
- ▶ Be especially cautious around horses at feeding time. To avoid being kicked, know exactly where the horses are before you turn your back or bend over to set out feed.

There's an old saying that goes, "If there's anything a horse can possibly get hurt on, you can bet he'll find it." Nursing and rehabilitation of injuries cost money in vet bills and take precious time away from training and riding. An injury is hard on both you and your horse, so run a tight ship and always be on the lookout for anything that might injure you or your horse or possibly cause a fire.

Make sure tools are put away and that carts and trunks don't block passageways. Loose items, such as rags, ropes, cat feed dishes, and old horseshoes, should be kept clear of horse traffic areas. Anything that your horse might step on or you might trip on could cause an accident. Wet spots on floors should be cleaned up and dried as soon as possible.

A good time to check facilities is when you're doing daily chores. Play detective and see if you can find the broken board, protruding nail, or twine on the ground that will ultimately save you time and money. Carry a small notebook and pencil to jot things down that you can't tend to immediately. It's good insurance to have someone check on the horses periodically throughout the day. If a horse gets caught in a fence or cast in a stall, the situation can quickly escalate from a minor problem to a life-threatening emergency in a matter of hours or even minutes.

Dogs can be wonderful, but they have no business around horses. Even when they have no designs on a horse, their sudden movements and the inborn "predator-prey" relationship between canines and equines are frightening to many horses. Keep dogs away from the barn and pens. Enforcing a "No Loose Dogs" policy will eliminate one common safety hazard for you, the horse, and the dog.

Check tack and equipment before each use. If a halter or rope is worn to the point where it might break, or an electric cord is nicked so you can see the wire inside, then they are unsafe and should be replaced.

25-Point Barn Safety Check

1. Be on a constant lookout for nails, metal, wire, twine, and other objects on the ground that could puncture a horse's foot or tangle his legs.
2. Check fences and pens for loose hinges, protruding nails, splinters, sharp edges, broken latches; put safety chains on gates with questionable latches (see page 54).
3. Make sure glass windows are protected from horse contact by heavy screens or metal bars.
4. Check electric fence wires and controllers often to make sure they're working.
5. Install adequate lights; keep light fixtures out of a horse's reach and use covered fixtures that protect the bulbs.
6. Make sure horses always have fresh water and that automatic waterers are functioning properly.
7. Keep grain stored in tightly sealed containers where a horse positively cannot get at it.
8. Be sure feeders are securely mounted with no sharp edges or protruding bolts.
9. Arrange halters so horses can't reach them but so they're easy for you to grab.
10. Protect switches and outlets from dust and moisture; make sure all wiring that a horse could reach is inside heavy conduit.
11. Keep floors clean, level, and in good repair.
12. Have fully charged A:B:C fire extinguishers clearly marked and easily accessible.
13. Store rakes, shovels, and brooms away from traffic areas on secure hangers.
14. Keep aisles clear by storing carts and trunks out of the way.
15. Keep medications, fly sprays, and such in a cabinet out of the reach of children, horses, and pets.
16. Unplug extension cords, clippers, and heaters when not in use; coil cords neatly on hangers. Replace or repair damaged cords. Don't use electric appliances near water.
17. Make sure paths to doorways are uncluttered and that the doors can open fully.
18. Keep approaches to the barn, especially concrete, free of ice and snow.
19. Illuminate pathways to the barn with lights controlled by a motion detector, a light detector, or a manual switch.
20. Have both horse and human first-aid kits fully stocked, easily accessible, and clearly marked.
21. Consider a surveillance camera to monitor activity and minimize theft and abuse.
22. Make sure loft railing and ladders are in good repair.
23. Protect hay drop holes with a trap door or railing; always look below and shout a warning before dropping hay.
24. Have an engineer or county inspector examine the roof for snow-load capacity; during heavy snows, pull snow off the roof with a T-board attached to a long pole.
25. Protect power pole guy wires with several wood posts to prevent horses from running into them.

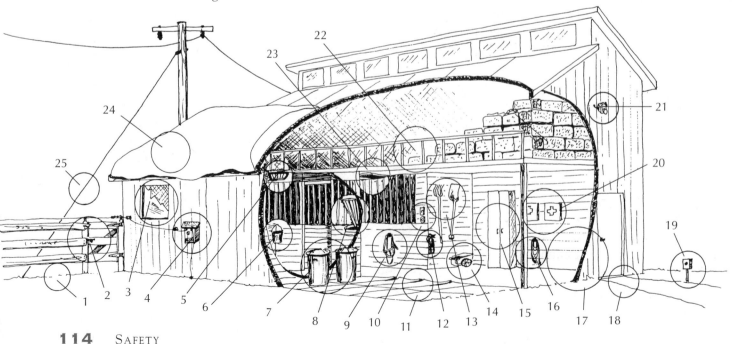

Facilities Safety

14.1 PERIMETER FENCE

Loose horses that leave your property can cause damage and traffic accidents that could prove to be your liability. Install a secure perimeter fence around the portion of the property where horses are contained. If a horse gets out of a pen or the barn, the perimeter fence will prevent him from leaving the property. Make sure all gates in your perimeter fence are closed at all times.

14.2 DANGEROUS BOLTS

These hinge bolts are very dangerous. A horse could cut a leg while rolling or hook his jaw or halter when sticking his head between the rails.

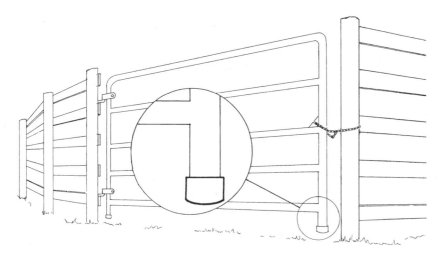

GATE CAPS

Many horses have injured their legs on the bottom corners of pipe gates. PVC end caps that fit snugly on the pipes can be secured with construction adhesive.

14.3 SAFE BOLTS

Cut bolts off flush with the nut to prevent injuries.

14.4 ANTIRUB STRATEGIES

Feeders almost always have an edge or corner that horses can rub on and ruin their manes and tails. Here, a plastic barrel is used to prevent a horse from rubbing on the bottom edge of the feeder. The barrel is secured to the metal panel with barbless fence wire. Instead of a barrel, the space could be filled with masonry or wood.

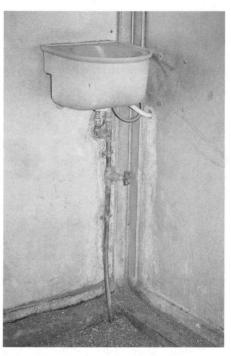

14.5 EXPOSED PLUMBING

Exposed pipes or valves, as shown here, can catch a halter or blanket, injure horses, or be damaged by kicking and chewing. Pipes and wire should be located inside the wall during barn construction, or enclosed in protective steel or heavy plastic pipe. The protective pipe should be set tight against the wall so a horse or foal cannot get a foot caught between the pipe and the wall.

Fire Prevention

There are several horse barn fires every day in the United States. Fires account for more horse deaths than all other disasters combined. Prevention is the best insurance against fire. Because of the risk involved, you can't be too careful, and your barn can never be too safe.

In spite of all precautions, fires will still occur from unpredictable causes such as lightning. Firefighters have a saying: "You have the first few seconds; the next few hours belong to the fire." The sooner you know about a fire, the faster you can act to save your horses and your barn. A home smoke detector doesn't work well in a barn, because dust and condensation can set it off falsely or clog it so it doesn't work at all. A better type of barn fire detector is a rate-compensating thermal detector that monitors changes in heat. To be sure the detector is noticed when it goes off, connect it to an outside alarm, to your house, and/or to a monitoring service that will call the fire department while you're away.

Fires are classified A, B, or C according to what is burning; a fire extinguisher is likewise rated according to the type of fire it is designed to extinguish. Class A fires involve solids such as paper, wood, and hay. Class B fires involve combustible liquids such as hoof black, hair polish, and paint. Class C fires are class A or B fires that also involve electrical wiring or equipment that is plugged in, such as vacuums, clippers, and heaters. A fire extinguisher rated A:B:C is best for a barn, because it will work on all three types of fire. For example, an extinguisher rated 3A:40B:C might weigh 8 to 10 pounds (3.6 to 4.5 kg), which is light enough for most people to handle. Units of this size will project a stream of chemical about 15 feet (4.6 m) for 8 to 15 seconds. Have one fire extinguisher this size or larger for every 3,000 square feet (278.7 m²) of barn. Locate one extinguisher in plain view at each door and have enough units so a person doesn't have to travel more than 50 feet (15.2 m) from anywhere in the barn to reach one.

Fire Prevention Checklist

1. Store the majority of your hay in a building separate from the barn.
2. Mow grass and weeds for at least 20 feet (6.1 m) around the barn, pens, and storage buildings (see chapter 16); keep free of brush and debris.
3. Post "No Smoking" signs and enforce them.
4. Keep the barn clean; remove combustible trash.
5. Clearly post phone numbers of the fire department and police next to phones, along with your address and directions to your barn.
6. Make sure electrical wiring complies with the National Electrical Code. Run wire in conduit; protect outlets and switch boxes from dust and moisture with weatherproof covers and replace broken faceplates; keep panel boxes covered.
7. Maintain a conscientious rodent control program. Rodents can chew through the plastic insulation on wires, causing them to spark.
8. Install fully charged fire extinguishers in plain view; clean and inspect once a month.
9. Keep a garden hose next to each hydrant; in freezing weather, make sure the hose is drained and free of ice (see Hoses, chapter 13). Buckets of sand are also useful for smothering a fire.
10. Install appropriate fire detectors inside the barn and hook them to a siren or bell on the outside of the barn.
11. Install a system of 12-inch-tall (30.5-cm) lightning rods, no more than 20 feet (6.1 m) apart, connected by a braided aluminum or copper cable to at least two 10-foot-long (3-m) ground rods.
12. Keep dust and moisture off light bulbs with protective covers.
13. Consider installing a sprinkler system.
14. Remove cobwebs and dust periodically.

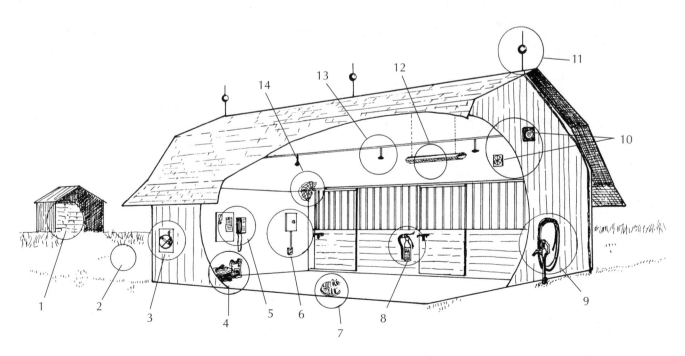

Fire Prevention

14.6 POSITIVELY NO SMOKING!
A "No Smoking" policy should be strictly enforced, and proper fire extinguishers should be close at hand and easily visible.

14.8 PLUG CAPS
These inexpensive plastic caps plug into standard outlets to keep them clean when they're not in use.

14.7 MAINTAIN SWITCHES
Dust and insects in switches are a fire hazard that can easily be avoided. The top switch, which remotely controls a yard light, needs a cover plate, and the switch below has a broken cover that should be replaced.

SPRINKLERS

At the onset of a fire, an automatic sprinkler system can buy you precious minutes needed to evacuate the barn. If your barn is connected to a municipal water system, you probably have adequate water pressure to run a sprinkler system. Sprinkler systems require a lot of water, however, and most rural stables are on domestic wells, which couldn't keep up with even one sprinkler head.

Safe Lights

Light bulbs break easily and can cause fire and/or injury. The best way to protect light fixtures is to keep them out of a horse's reach. Try to mount lights 11 feet (3.4 m) or higher and never mount them lower than 9 feet (2.7 m). Fixtures mounted lower than 11 feet (3.4 m) should be protected by a heavy metal grille-like guard that encases the bulb and is securely attached to the fixture or to the mounting surface.

Clean Floors

Loose twine in the barn can cause you to trip and be injured. Loose twine in pens can become tangled around a horse's feet and cause serious injuries. Make it a habit to bundle up all twine and wire and throw it into an empty feed bag near the hay area. Keep aisles as uncluttered as possible so there's plenty of room to safely lead a horse or work on him in the aisle. Sweep up hay dregs daily and feed them. Loose hay can be very slippery to walk on and is a fire hazard as well.

14.9 **Don't underestimate the power or reach of a horse. It can rear up and easily break through wire basket–type light protectors. Even if out of a horse's reach, this basket is totally ineffective in keeping dust from accumulating on the bulb. Also, the electrical wire should be enclosed in conduit.**

Security

Theft of tack or horses can be a devastating experience. An entrance gate that can be securely locked discourages theft, and may also help keep out solicitors looking to steal some of your precious horse time. Make sure the gate is installed so it can't be lifted off the hinges. Alternatively, secure the hinged side with a chain and padlock.

A tack room with secure door locks and no windows can minimize theft. Cover existing windows with sturdy steel grilles. Mark all tack with an ID number, and keep tack out of sight when you're not using it, and under lock and key when you're gone. Make a written and photographic inventory of your tack. Check with your insurance company to see if certain types of locks are required to validate theft coverage. If you keep your barn clean and organized, you'll be more likely to notice when things are missing or out of place. Keep several copies of current photos, and written descriptions, of all horses in the stable for immediate distribution to police, local veterinarians and farriers, and others.

"Watch animals," like dogs or geese (which are less inclined to take bribes than dogs are), can alert you to strangers. Yard lights that come on at dusk or are triggered by motion sensors can be a great deterrent. Horses shouldn't be turned out wearing halters for safety reasons, but also because horses wearing halters are easier for thieves to catch.

Surveillance cameras can allow you to keep an eye on areas such as a tack room, foaling stall, barn aisle, or main entry from your house or office building on the premises. Just the sight of a camera can be enough to discourage a thief or a vandal.

15

✦ EMERGENCY ✦

Emergencies, whether caused by accidents or natural disasters, are inevitable and unpredictable. Proper preparation can minimize damage to your facilities and reduce the risk of injury and discomfort to you and your horses.

General Preparation

Preparation includes having emergency supplies on hand, knowing what to do and what not to do during and immediately after a disaster or emergency, and practicing emergency procedures.

▸ Learn basic first aid for people and horses.

▸ Know what to do, and what not to do.

▸ Initiate a specific plan for the types of emergencies most likely to occur in your area, such as floods, hurricanes, or tornadoes, and practice the plans with everyone involved in the operation of your barn.

▸ Know who to call, and when to call. Don't tie up phone lines needlessly. During a natural disaster, call 911 and emergency numbers only in life-and-death situations.

▸ Have phone numbers for sheriff, police, fire department, vet, and neighbors clearly posted and on clearly marked speed-dial buttons, along with 911, if possible.

▸ Have the physical address of your barn and detailed directions telling how to get there posted by all phones.

▸ Post your house number where your driveway joins the main road, so it is clearly visible from both directions; keep weeds, brush, and trees cut back so they don't block the sign.

▸ Keep extra halters and leads in your house so you can more quickly catch loose horses without going to the barn.

▸ Keep a two-week supply of feed for horses in waterproof containers.

▸ Keep a three-day supply of water for each of your horses.

▸ Know where and how to shut off electricity, gas, and water.

▸ Arrange for alternative sources of electricity, water, light, and heat in case a of power outage.

SERVICE SHUTOFF TIPS

▸ Know where the main electric service panel is and know how to disconnect it. Make sure it isn't covered by tack and that there's a clear path to it at all times.

▸ Know where the main water valves are and how to shut them off in case of a waterline break, an automatic waterer malfunction, a flood, or an earthquake.

Injury to Person or Horse

If you, another person, or a horse gets injured, there's a good chance you'll need the help of another person, and fast. A two-way personal pager and an intercom system between your barn and other buildings are two of the quickest means of contacting someone you know. A phone in the barn can connect you with 911 and other emergency providers.

You can use intercoms that operate by FM signals through the existing electrical wiring on your property, as long as your buildings are all supplied by the same electrical meter (see appendix). Just plug the intercoms into standard outlets, push a button, and talk. Some intercoms have a "lock on" feature that allows you to continuously monitor the sounds in a room, such as a foaling stall or tack room, for example.

Horse Injury Action Plan

▶ Remain calm.
▶ Have the person who is most experienced with horse first-aid deal with the injured horse.
▶ Keep the horse as calm, quiet, and comfortable as possible; covering a horse's eyes often helps calm him down.
▶ Protect yourself and others from injury. Only restrain the horse if you can do so safely; if the horse is on the ground, hold his head down to prevent him from getting up.
▶ Call a veterinarian if necessary, describing the injury and state of the horse as accurately as possible.
▶ Take and record the horse's vital signs.
▶ If the horse is not yours, notify the owner. If the horse is insured, it may be necessary to contact the insurance company if surgery or euthanasia is required.

911 EMERGENCY NUMBER

The 911 dispatch is available 24 hours a day for police, fire, and emergency medical services. However, 911 is *only* a dispatcher. Quick response depends on vital information you provide. An operator will listen to your information and then send the appropriate police, fire, or emergency medical team located in your area. Do *not* call 911 for information such as road or weather conditions.

What to Do

1. Dial or press the numbers 9-1-1 or use a preprogrammed speed-dial button on your phone.
2. Try to stay calm, and clearly state the nature of your emergency (e.g., fire, power line down, type of injury to person or horse).
3. Give your street address.
4. If necessary, give directions to your farm, reading off a card (posted by each of your phones) to ensure accuracy.
5. Answer other questions the dispatcher might have.
6. Don't hang up until the dispatcher does.

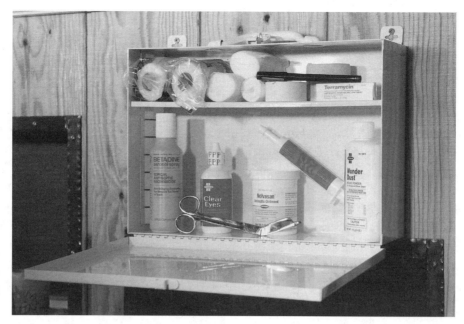

15.1 Keep a well-supplied horse first-aid kit ready in a convenient place.

SUGGESTIONS FOR YOUR HORSE FIRST-AID KIT AND CUPBOARD

General Items
- Band-Aids, assorted sizes
- Sterile gauze pads
- Adhesive tape
- Conforming gauze roll
- Disposable diapers
- Elastic bandage
- Crepe bandage
- Thermometer
- Disposable syringes and needles
- Disposable razors
- Safety pins (assorted sizes)
- Instant cold compress
- Protective boot
- Weight and height tape
- Lubricating jelly
- Current first-aid manual
- Rubber gloves
- Towels (cloth and paper)
- Phenylbutazone (anti-inflammatory to relieve pain and swelling; tablets or paste)

Instruments
- Chain
- Twitch
- Watch
- Stethoscope
- Heavy-duty scissors
- Forceps
- Sharp pocket knife
- Leatherman (Leatherman Tool Group, Inc. Portland, OR). Multipurpose stainless steel folding tool that contains a knife, pliers, screwdriver, and other common tools or similar compact multitool
- Wooden sticks for applying ointments
- Flashlights, two
- Fresh spare batteries

Refrigerated Items
- Liquid antibiotics
- Epinephrine (antidote for anaphylactic shock caused by allergy to penicillin, bee stings, and chemicals)

Natural Disasters

Items to have on hand:

▸ Bleach, unscented, for treating contaminated water
▸ Plastic garbage bags for lining water containers and storing tack, feed, and bedding

The following procedures and tips can apply to most natural disasters:

▸ Stay calm.
▸ Listen for broadcast emergency instructions on a car radio or battery-powered radio.
▸ Don't use your telephone, except in an extreme (life-threatening) emergency.
▸ Stay at least 33 feet (10 meters) from downed power lines.
▸ Never mix bleach with ammonia when cleaning up your barn after a disaster; the fumes produced are toxic.
▸ Check your neighbors after looking after your own horses.
▸ Confine frightened horses and dogs.
▸ Contact your utility company or police department to report downed power lines or broken water or gas lines. If you see sparks or frayed wires, or smell hot insulation, turn off the electricity at the main fuse box unless you have to step in water to do so.

Power Loss

Electricity is the lifeblood of your barn. It provides power for three necessities: light, heat, and water. Power can be interrupted for days or weeks at a time by wind and ice storms and other unpredictable events. Keep your barn operating with an alternate electric source or alternatives to electric lights, water, and heat.

Gas-Powered Generators

The cost of a generator mainly depends on the output, which is rated in watts. Depending on the output of the generator, you might have to take turns between running a water pump and a heater or lights.

Because of exhaust fumes and noise, the generator should only be operated outside the barn, so make sure you have enough heavy-duty extension cords with the right plug configurations to connect the generator to the water pump, heaters, and lights. Extension cords should be 12 gauge or heavier. (As the gauge gets higher the wire size gets smaller, so a 14-gauge extension cord is *smaller* or *lighter* than a 12-gauge cord.) The gauge of the wire used in an extension cord is usually imprinted on the sheathing of the cord.

If you are not on municipal waterlines, but have your own well, check to see if your pump is permanently connected ("hardwired") or if it plugs in to an outlet. If it's hardwired, have it modified to plug in, to ensure you'll be able to connect it to a generator.

DETERMINING REQUIRED WATTS

To find how many watts a motor requires, multiply its voltage by its amp rating (stamped on a metal plate fastened to the motor). A 120-volt water pump rated at 10 amps would use 1,200 watts (120 x 10) of electricity. Add the watt requirement of your essentials to find the size generator you'd need.

15.2 **Gas-powered generator**

Fuel

Gasoline-powered generators will typically run from 2 to 8 hours on a gallon of fuel. Be sure to keep fresh fuel on hand to run your generator. Keep the tank 95 percent full to prevent condensation and still allow for expansion. Gasoline oxidizes and goes stale in 3 to 6 months, making a sludge that can plug injectors and fuel lines. Gas oxygenated with ethanol goes stale more quickly; premium grades stay fresh longer. Adding a gas stabilizer can help.

Water Supply

On average, a horse needs 10 gallons (37.9 L) of water each day. You should plan to have access to at least a 3-day supply (30 gallons [113.5 L] per horse) in case of power failure or contamination of your water supply. Make it a practice to keep all of your larger water tubs and troughs filled with fresh water and you might not have to worry about water for the horses during a power outage. If you have a natural source of uncontaminated water, such as a pond or creek, you can use it temporarily until the power is restored. A small fenced-in area with safe access to the water's edge can be used to turn out each horse for a drink. If your natural water source is not enclosed, the horses can be led to the water twice a day, but this can be pretty labor intensive and, as you know, you can't *make* a horse drink. Horses tend to drink more readily when *left* with water rather than when *led* to water.

Other alternatives include:

▸ Installing a hand pump on your well
▸ Installing a gravity feed waterline from a natural water source or a cistern full of stored water
▸ Storing water in buckets, barrels, garbage cans, or even boxes lined with plastic garbage bags.

Treating Contaminated Water

Contaminated water can be purified for human use with chlorine bleach or iodine. To be fully effective, purification compounds must be in contact with the water for at least 30 minutes to kill all bacteria present. The water must be well mixed and should have a slight iodine or chlorine taste. Check with your doctor or local health authorities for more complete information.

▸ **Boiling.** If the water is clear, boil it for 10 minutes. If water is cloudy, filter it by pouring it through a coffee filter and then boil it for 10 minutes.
▸ **Chlorine bleach.** If the water is clear, add 8 drops per gallon (3.8 L). If the water is cloudy, add 16 drops per gallon (3.8 L). Shake and let stand for 30 minutes before using. A slight chlorine odor should be detectable in the water. If not, repeat the dosage and let stand for an additional 15 minutes.
▸ **Iodine.** Tincure of iodine or iodine-based tablets can be used to kill hard-shell protozoans such as Giardia and Cryptosporidium that are not affected by bleach. Confer with your local health department if you suspect this problem.

> ### TIP FOR STORING WATER
>
> Use clean glass or heavy plastic containers for long-term water storage. To ensure the water stays disinfected, treat it with liquid chlorine bleach (unscented and containing 4 to 6% sodium hypochlorite). Use 16 drops per gallon, or 1 teaspoon (4.9 mL) per 5 gallons (1.9 L). Fill containers to the top and keep them tightly sealed. Replace stored water every 6 months.

Water Supply

CISTERN WATER STORAGE

A cistern (a plastic or concrete water storage tank), located uphill from the barn, can provide water to the barn by gravity feed in the event of a power failure and as an added source of water in case of fire. In cold climates a cistern should be buried so the bottom is below frost line, with at least 1 foot (.3 m) of earth over the top to keep the water from freezing. In areas where the temperature generally stays above 32°F (0°C), a cistern can be aboveground. The bottom of the cistern must be above the level of the water hydrants and faucets at the barn in order for the water to gravity feed. The greater the altitude difference between the cistern and a

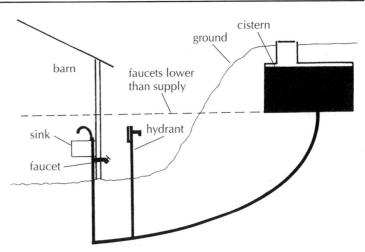

faucet, the greater the water pressure will be coming out of the faucet. Have cistern water tested by the county health department at least once a year to make sure it's safe.

15.3 WATER STORAGE BARREL

Barrels can also be used for storing an emergency water supply, although this is less practical in cold climates where the water would need to be kept from freezing. At 10 gallons per day, one 50-gallon (189.3-L) barrel of water would last one horse 5 days. Plastic barrels, such as those used in the food industry, are better than steel drums because they won't rust. Check to be sure the barrels were not previously used for harmful substances. Scrub the barrels very thoroughly before using them for water so you don't inadvertently cause your horses to go off water due to a strange taste.

15.4 WATER STORAGE BUCKETS

Fifteen gallons (56.8 L) of water stored in these three 5-gallon (18.9-L) buckets would last one horse 2 days at the most. In the meantime, a board across the tops of the buckets makes a useful shelf in the tack room.

Lights

Battery-powered emergency lights use either standard incandescent bulbs or energy-saving fluorescent bulbs. Place several flashlights and/or lanterns at strategic locations in your barn. For example, keep a flashlight by each entry door near the light switch where it can be easily grabbed, even in the dark. A lantern, which lights a broad area and leaves your hands free, would be appropriate in the feed room, either sitting on a shelf or hanging from a hook. Some flashlights clip to your clothing; the ultimate hands-free light is a headlamp.

Lanterns that burn kerosene or a similar fuel use a cotton wick that must be trimmed and adjusted properly for maximum light and minimum smoke. Propane lanterns and certain other liquid-fueled lanterns use a fabric mantle, which burns much brighter and cleaner than a wick. We've all seen at least one old western movie in which a barn fire is started by a kerosene lantern. Although modern liquid-fueled lanterns generally are much safer than the old-style kerosene lanterns, it still is a good idea to evaluate each lantern's design and safety features closely when considering it for barn use.

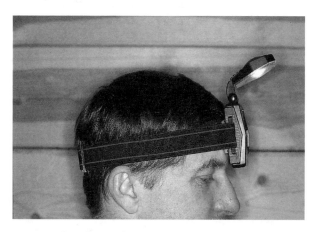

15.5 A headlamp like this one is ideal for a horse person taking care of emergency measures or doing chores in the dark. It shines directly in front of your face, leaving both your hands free. If you have cold winters, make sure the kind of headlamp you choose can be worn with your winter hat.

Heat Source

Portable propane or kerosene heaters are available in a wide variety of styles and sizes and are lightweight enough to easily move to where they are needed. They use oxygen and give off carbon monoxide and other gases, however, and most are not recommended for closed occupied spaces. Heaters that can be used in living spaces usually have a vent that carries exhaust gases to the outside, which limits their portability. Unless the heater is specifically made for indoor use, be careful about using it in a room occupied by you or your animals.

Heat output is measured in British thermal units, or BTUs. The smallest propane heaters screw onto a disposable propane cylinder and put out 3,000 to 4,000 BTUs, which would be good for short-term heating of a small area (thawing out a frozen water pipe, for example). Larger propane

15.6 This pair of 500-watt halogen lights on a stand can be placed wherever light is needed and plugged into a generator.

heaters attach either directly, or by a hose and coupling, to a 20-pound (9.1-kg) or larger propane cylinder and put out from 10,000 to more than 80,000 BTUs, which would be suitable for heating a room. An 8-foot by 10-foot (2.4-m by 3.0-m) insulated tack room would require a 5,000-BTU heater to keep it comfortable during subfreezing weather. Rental stores can provide a propane cylinder and a heater with enough output for a tack room.

Propane heaters typically burn 1 to 2 pounds (0.5 to 0.9 kg) of fuel per hour. Propane is sold by the gallon but the cylinders are filled by weight. There is an approved weight stamped on each propane cylinder. The empty cylinder is placed on a scale and filled with propane until it reaches this approved weight. One gallon (3.8 L) of propane weighs 4.24 pounds (1.9 kg), so a 20-pound (9.1-kg) cylinder will hold about 5 gallons (18.9 L) of propane.

Kerosene is a fuel similar to gasoline but more oily. It is usually sold at ranch supply stores, full-service gas stations, and oil distributors, but it's not readily available in many areas. When buying kerosene you must usually provide your own container, and it must be approved for kerosene. A kerosene heater is filled by carefully pouring the fuel into a reservoir on the heater. Kerosene has a strong smell that will linger if spilled. Some kerosene heaters have safety features designed to prevent fuel leakage if the heater is tipped over. A kerosene heater that puts out 23,000 BTUs can run up to 8 hours on a gallon (3.8 L) of fuel.

Evacuation

If there's ever a time you need your horses to load in a trailer, it's when you need to evacuate them in a hurry. Train your horses thoroughly. Practice loading your horses at night and during rainstorms, and then practice some more.

Your evacuation plans for floods or hurricanes should include enough time to make several trips to move all your horses, if necessary, and arrangements for a place to keep your horses until you can return home. Become acquainted with your horsey

15.7 This propane heater quickly screws onto a 20-pound (9.1-kg) tank, such as is used with barbecue grills, and doesn't take much storage space when not needed.

neighbors as far down the road as you can, and make a list of neighbors and facilities that will act as holding pens in case of an emergency.

If you think there's any chance you'll need to evacuate, bring your horses in off pasture and pack your trailer. Be sure to keep up on your truck and trailer maintenance so they are ready to haul horses at a moment's notice. Keep tires repaired and inflated and hitches and lights in good working order. (See *Trailering Your Horse* for more information on trailer maintenance and safety.)

Listen to a battery-operated radio for storm updates and if evacuation is advised, leave early. In hurricane country, plan to evacuate your horses 1 or 2 days before the storm is scheduled to hit, in order to avoid traffic and high winds while towing your trailer. In flood country, evacuation is much simpler and safer when done before floodwaters become too deep to drive through.

Follow recommended evacuation routes, because your normal shortcuts may be blocked. Take extra care when driving, because familiar roads may be dramatically different when flooded. Be alert for damaged bridges, mud slides, washouts, and downed power lines. Look for emergency personnel and information signs. Obey officials who may be directing traffic and other operations related to the flood; they are there to help you. Do not drive through a flooded area, even if it looks shallow enough to cross; roads concealed by water may not be intact, and 2 feet (0.6 m) of moving water can easily sweep a pickup and horse trailer away. Most deaths due to flash flooding occur when people drive through flooded areas.

ITEMS TO TAKE WITH YOU

Many small items can be packed in 5-gallon (18.9-L) plastic buckets with tight-fitting lids.

- ❏ Horse ID papers
- ❏ Personal identification (you may need it to reenter your neighborhood after the storm)
- ❏ Money and/or charge cards
- ❏ Necessary medication for people and horses
- ❏ First-aid kit
- ❏ Plastic garbage bags
- ❏ Premoistened packaged towelettes
- ❏ Warm clothing, including waterproof outer garments and footwear
- ❏ Sleeping bags or blankets
- ❏ Personal care items
- ❏ Battery-operated radio
- ❏ Flashlights and fresh spare batteries
- ❏ Important personal and family documents
- ❏ Feed and water for horses
- ❏ Saddles, bridles, halters, and lead ropes in case you need to abandon your vehicle

Before You Leave

Turn off the electricity, gas, and main water valves and unplug your appliances. If time allows, move tack and grain to a high and dry place. Call a neighbor, friend, or relative or leave a note in a plastic bag tacked to the door telling where you are going and when you left. Include a contact phone number if possible. Make sure you lock all barn doors and windows.

Fire

Some disasters, such as tornadoes, floods, or hurricanes, give at least some warning, but a fire usually happens instantaneously. Horses are instinctively frightened of fire and smoke, so the better trained they are to lead and the more practice you've had handling them, the better able you'll be to get them safely out of a burning barn without panicking. If the barn is already on fire, soak a towel and tuck it under the crownpiece of the halter so it covers the horse's eyes and nose. Remove synthetic tack, such as nylon blankets, that could melt onto the horse.

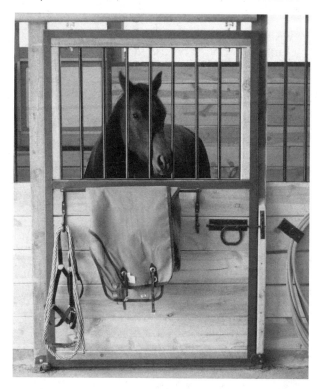

15.8 Have a halter hanging by every stall so you can move horses quickly and safely.

Extinguisher Practice

Everyone associated with your stable should practice operating one of your fire extinguishers, even if not on a fire. Have the unit recharged after practice so it's ready to go.

1. Pull out the safety pin.
2. Stand 7 to 10 feet (2.1 to 3.0 m) from the fire.
3. Aim at the base of the fire.
4. Squeeze the trigger.
5. Spray with a back-and-forth motion across the fire's base.

Be Prepared

- ❏ Clearly post phone numbers for 911, fire, sheriff, and neighbors; program speed-dial buttons, if possible. Post your barn's exact physical address and accurate driving directions.
- ❏ Equip your barn with a sufficient number of fire extinguishers.
- ❏ Have a properly fitted halter hanging near each stall.
- ❏ Practice leading every horse out of all barn exits.
- ❏ Make sure doors are unobstructed and latches are operable.
- ❏ Make sure there's a fenced area away from the barn where evacuated horses can be contained.
- ❏ Invite a representative of your local fire department to visit your barn to become familiar with the layout and to advise on modifications to make it more accessible to firefighting equipment.
- ❏ If you keep your front gate locked, leave a key with your local fire department.

FIRE ACTION PLAN

1. Get people out of the barn.
2. Call 911 and/or the fire department; be sure they know what kind of fire and how to get there.
3. Get horses out of the barn and out of pens next to the barn; put them in a distant pen. If you can't lead the horses out, drive them out and make sure they can't get back inside. This is a situation in which you must have a perimeter fence and a securely closed entrance gate to keep the loose horses on your property.
4. Have someone unlock the front gate, move vehicles if necessary to make way, meet the fire crew, and direct them to the fire.
5. Fight the fire with fire extinguishers and/or hoses if you can do so safely. Soak the unburned walls, bedding, and hay to keep the fire from spreading.
6. Keep dogs contained and stay out of the way for the fire crew's arrival.

FIREFIGHTING TIP

Water can be useful for fighting Class A fires (hay, wood, paper, cloth, plastic, rubber). Do *not* use water on Class B fires (grease, gas, oil, solvents), because water can spread the fire, making it worse. Also, do *not* use water on Class C fires (any fire involving electric wiring or plugged-in tools and appliances), because you could be electrocuted. Sand can be used to smother any kind of fire; keeping a bucket of sand near each entrance is a good precaution.

Wind and Storms

During high winds, horses in a solid barn are better protected from blowing debris than those outside. On the other hand, horses might be safer outdoors than inside a structure that might collapse. It's often a tough call to make and there are no hard-and-fast rules.

If you turn your horse out during a storm, a well-fitted fly mask can protect his eyes from wind-blown debris. Make sure every horse is identified with your name and phone number on a breakaway halter or with a plastic neckband (available through vet catalogs), or mark them with grease crayons (used for marking cattle and sheep) or even spray paint. A breakaway halter would allow rescuers to more easily catch your horse, yet would minimize the chance of injury if caught on something. Permanent identification such as tattoos, brands, and microchips are not as easily recognizable to rescuers, but are better protection against thieves who take advantage of disasters.

TIPS TO PROTECT AGAINST WIND DAMAGE

- Trim dead and weak tree limbs and remove dead and weak trees around barn and pens.
- Tie down or move inside all objects that might be moved by strong winds, such as barrels, jumps, wooden pallets, mounting blocks, plywood, lawn furniture, toys, hay tarps.
- Close all doors, windows, and shutters if the barn has them.
- Make sure water tubs are filled to keep them from blowing away.
- If you turn horses out, close barn doors so they can't run back inside.
- Secure horse trailers with mobile-home tie-downs.
- Turn off gas supply to prevent fire from broken fittings and lines.

Floods

Floods are the number one weather-related killer in the United States. Although some floods are slow in coming, flash floods usually give no warning. Floods can result from excessive rainfall, earthquakes, tidal waves, hurricanes, or dams that give way. In some cases, barns and pens may be protected by using sandbags or polyethylene barriers. This precaution requires specific instructions that can be obtained from your local emergency officials. Horses perish not only from drowning but also by getting stuck in mud, so if you turn them out make sure it's on high ground. In some cases, you may need to evacuate your horses.

When There Is Immediate Danger of Flooding

- Turn horses out on high ground or evacuate them.
- Shut off power to the premises. If the main switch is in an area that is already wet, stand on a dry board and use a dry stick to turn switch off. Do not attempt to turn off power if the room is already flooded.
- Turn off water supply.
- Disconnect all electrical appliances and, if possible, move them to a higher level. Thermally insulated appliances such as freezers, refrigerators, and ranges should be given priority. Any appliances that cannot be moved should have motors, fans, pumps, and so on removed to higher levels if there is time.
- To prevent oil and water tanks from floating away, fill them if possible, then plug vent holes. If unable to fill them, weight them down with sandbags or rocks or wedge them against a solid object.
- A propane tank will float whether full or empty, so turn off the valve, disconnect tubing, and plug it. Tie a rope, chain, or cable around the tank and anchor it to prevent it from floating away.

- Plug all sewage connections in the barn (i.e., toilet, sinks, showers) with wooden plugs or other devices. To keep plugs in place, weight them down or use boards to brace them to walls or ceilings.
- Remove or move to higher levels items that could be damaged by flooding or those that may float and cause damage or pollution, such as tack, tools, feed, pesticides, weed killers, and fertilizers.

After the Flood

- To prevent water-damaged papers from further deterioration, store them in a freezer until they are needed.
- Record details of flood damage, by photograph or video if possible. Register the amount of damage to your barn with both your insurance agent and local municipality as soon as possible.
- Rinse all floors and walls, then clean as soon as possible with a 10% bleach solution (9 parts water to 1 part chlorine bleach), with nonammonia detergent added as needed. Dry thoroughly.
- Treat any areas of mold, which can lead to serious health problems, with chlorine bleach.

CAUTION

Never mix chlorine bleach and ammonia; together they produce deadly chlorine gas.

- Ventilate and/or dehumidify the barn until it is completely dry.
- Keep horses well away from downed power lines. Electrical currents can travel through the water for more than 100 yards (91.4 m).
- If your well has been flooded, assume the water has been contaminated. If you are

on a public water system, listen to your radio and television for news from public health departments to find out if your water is contaminated (see Treating Contaminated Water, page 124).
- Watch out for animals, especially poisonous snakes, that may have come into your barn with the flood waters. Use a stick to lift debris before handling it.

Extreme Heat

Extreme heat, especially coupled with high humidity, can cause dehydration and death.

Signs of Heat Stress

- Dehydration
- Sweating
- Elevated respiration rate and shallow breathing
- Elevated heart rate
- Bulging veins
- Depression, lethargy
- Dark red or muddy mucus membranes

Treatment

- Get the horse into the shade.
- Remove all tack.
- Cool the horse quickly with fans and a hose or sponge, using the coldest water you can, even ice water.

Prevention

- Allow the horse to drink as much water as he wants, whenever he wants, even during exercise.
- Leave a wet horse unblanketed for maximum cooling.
- Provide shade for horses at all times.
- Do not overwork horses in extreme heat.
- Set up fans to blow on horses in stalls.

Extreme Cold

A horse can suffer cold stress and sustain permanent damage, such as frostbitten ears, if not provided with shelter from wind and moisture during extreme cold.

Signs of cold stress
- Depression, lethargy
- Shivering
- Very cold ears

Treatment
- Bring the horse in out of the wind.
- If he's wet, dry his large muscle masses vigorously with towels or burlap, but be gentle around his head and ears.
- Put a dry blanket on the horse.
- Walk the horse to get his circulation going.
- Provide access to radiant heat source.
- Feed plenty of grass hay.

Snow and Ice

Snow and ice can make it very difficult and dangerous to get around and care for your horses. Your first priority is to make paths to feed and water, so have shovels and snowblowers ready where you can get to them. A tractor equipped with a blade or loader can make clearing paths and cleaning pens much easier. Snow melting on the roof often refreezes on the edge of the roof and forms an ice dam that can cause roof leaks, especially on heated buildings. These mini-glaciers, along with accumulating snow, can come crashing down in large chunks, frightening horses and filling pens. Heat cables (see Appendix) can be attached to the roof edge to keep melted water running off the roof (see also chapter 16).

15.9 Healthy horses with a natural winter coat can easily tolerate temperatures below –20°F (–29°C) if allowed to stay dry and out of the wind. In fact, horses that are provided shelter will often choose to stand out in the elements.

TIPS FOR COLD-WEATHER CHORES

- Allow more time for chores to deal with frozen water.
- Dress carefully. Don't overdress or you might sweat and then get chilled.
- Wear loose-fitting, lightweight clothing in several layers that can be removed to avoid perspiration and subsequent chill.
- Wear a hat that covers your ears. Half your body heat is lost through the head.
- Wear mittens, snug at the wrist; they're warmer than gloves.
- Avoid overexertion, such as shoveling heavy snow, pushing a cart, or walking in deep snow. The strain from the cold and the hard labor may cause a heart attack. Sweating could lead to a chill and hypothermia.
- Cover your mouth to protect your lungs from extreme cold. A scarf or fleece "neck gaiter" keeps your neck warm and can be pulled up around your mouth and nose.
- Stay dry.

Thawing Frozen Pipes

There is nothing quite like dealing with a frozen water pipe in subzero temperatures to test a person's good nature. Start thawing at an open faucet nearest the frozen section by applying low-level heat using one of the methods listed below. If the faucet is frozen, thaw it out first and work your way from the faucet down the pipe. If steam is formed, the open faucet will allow it to escape and not burst the pipe. Also, the open faucet will tell you when the water begins to flow. Let the water run for several minutes to completely thaw and flush all of the ice pieces from the pipe.

There are several other thawing methods you can try:

▶ A household iron, heat lamp, propane torch, or hair dryer can be used to heat the pipe. Be very careful not to melt plastic pipe. Move the heat source back and forth and never heat a pipe any warmer than you can touch.
▶ If hot water is available, trickle it over towels or rags wrapped around the pipe.
▶ Apply heating cable to the frozen pipe. Adding insulation over the heating cable will speed the thawing process.

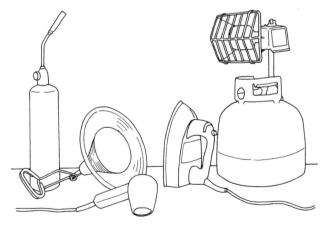

Items that can be used to thaw frozen pipes include hot water and towels, and from left to right: propane torch, heat lamp, hair dryer, electric iron, and propane heater.

▶ If the frozen section of pipe is inaccessible, you may be able to use an electric or gas heater to warm the air around the pipe enough to thaw it. If the heater doesn't have a built-in fan, use a separate fan to force air into the wall cavity or crawl space containing the pipe.
▶ Some electric welders can be used to thaw frozen metal pipes. Not all welders are the same, and it's important to check the welder manual to see if and at what settings the welder can be used for thawing pipes. This method can only be used on continuous sections of metal pipes. It won't work on plastic pipes or if there are plastic joints in the section you want to thaw. Disconnect any electric system ground wires attached to the pipes to prevent damaging the building's wiring system. With the welder off, attach the welder cables to the pipe on either side of the frozen section. When the welder is turned on, the electric current warms the pipe and thaws the ice.

HAVE YOUR TRACTOR READY

▶ Check antifreeze.
▶ Keep fuel tank filled with fresh fuel. (See page 124.)
▶ Keep tire chains in good repair and have them accessible and ready to put on.
▶ Have heavy-duty jumper cables on hand in case tractor won't start.
▶ Park tractor where it's not blocked by other vehicles so you can get it out or jump-start it with another vehicle if necessary.
▶ Have attachments such as a loader bucket and blade on the tractor ready to go, or store them where you can easily hook up to them in deep snow.

Earthquake

Go through your barn, imagining what could happen to each part of it if it were shaken violently. Check for barn hazards: Are the walls braced? Are roof trusses and rafters firmly attached? If hay is stored in the loft, will the floor withstand an earthquake? Consider storing hay in a separate building rather than in an overhead loft. Seek advice from professionals (insurance agencies, engineers, architects, builders) if you are unsure of what to do.

▶ Secure appliances, such as a water heater, that could topple and break gas or water lines.

▶ Secure top-heavy shelving units to prevent tipping. Keep heavy items on lower shelves.

▶ Hang tools such as rakes and shovels on hooks that hold them securely.

▶ Use safety latches on cupboards to keep contents from spilling out.

▶ Keep grain in containers with tightly closing lids.

▶ Keep flammable items and chemicals in a safe cupboard if they can't be stored in an outside shed.

During an Earthquake

▶ Get horses outside and keep them clear of buildings and wires that could fall on them.

▶ Remain in a protected place until the shaking stops. Anticipate aftershocks that may occur soon after the first quake.

▶ Try to remain calm; keep your horses calm.

▶ Move horses away from waterfront areas because of the threat of tidal waves.

▶ As soon as you can, fill as many water containers as you have with clean water.

16

✦ SEASONAL REMINDERS ✦

When you keep horses, there is never a shortage of things to do. In fact, it is easy to get so busy with daily routines that some seasonal items are forgotten. But if you get in the habit of making specific seasonal checklists, it will ensure that certain important things get done that otherwise would slip through the cracks. If you use one season to prepare for the requirements of the following season, your horses and facilities will always be ready and you can avoid falling into the eternal "catch up" syndrome. Winterizing your tractor in the fall, for example, means you won't be scrambling around looking for tire chains and antifreeze when your driveway is knee-deep in snow. Use the slow winter months to repair tack and you will have that much more "horse" time when spring weather beckons. If you keep your horses' feet regularly trimmed throughout the winter, you will likely avoid therapeutic farrier bills in the spring and your horse will be ready for an active season of training and riding. Take the time to do all of your hay shopping in the summer, when there is a large selection and hay is plentiful. In the throes of winter, when your hay barn is full and hay is scarce and expensive, you'll appreciate the money and time you saved by shopping early and stocking up. You can think of these seasonal preparations as "previews of coming attractions" that get you, your horses, and your stable ready for the next show that nature has in store.

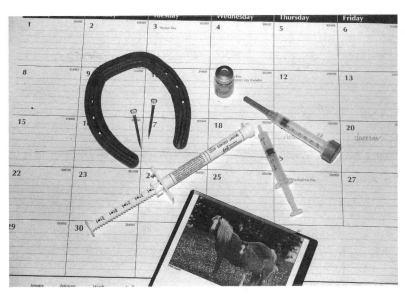

16.1 KEEP ON SCHEDULE

To keep track of tasks and appointments that are scheduled at regular intervals year-round, write them on a calendar. Include such things as:

▶ Hoof care (every 6 to 8 weeks)
▶ Deworming (every 8 weeks)
  (see Parasite Control, chapter 9)
▶ Cleaning and inspecting fire
  extinguishers (every 4 weeks)
▶ Vaccinations (as needed,
  depending on horse)

Year-Round Periodic Appointments

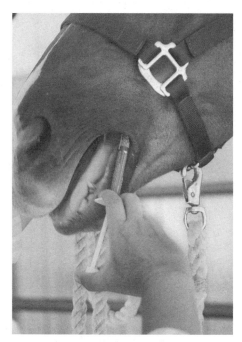

16.2 DEWORM REGULARLY
Most horses, unless they are on a continuous feed-through program, need to be dewormed every 2 months to protect them from the ravages of strongyles, bots, and other parasites.

16.3 KEEP HOOF CARE ON SCHEDULE
Horses need professional hoof care every 6 to 8 weeks, whether they are shod or barefoot, and whether they are in work or turned out. Avoid costly repairs or lameness by staying on schedule.

Fall

It seems that every year we get lulled into thinking the marvelous fall weather we're enjoying will stretch into spring — not so! Now is the time to prepare for Old Man Winter.

Checklist

- ❏ Remove bot eggs (see page 72).
- ❏ Give additional influenza vaccination if necessary.
- ❏ Wash fly sheets and store away.
- ❏ Check that winter blankets are ready to use.
- ❏ Check tractor antifreeze.
- ❏ Make sure tractor and generator have fresh gas (see page 124).
- ❏ Check or install heat tapes on pipes.
- ❏ Oil hinges and latches to prevent freezing.
- ❏ Drain water lines in unheated buildings to prevent pipes and faucets from bursting.
- ❏ Check that snow removal equipment is ready.
- ❏ Put snow pads on horses.
- ❏ Install snow fence.

Field Washing a Fly Sheet

Field washing on a concrete surface using a stiff broom is an excellent way to get a very dirty PVC sheet absolutely clean. This method can also be used on turnout sheets and winter blankets.

Alternatives are machine washing or using a high-pressure sprayer at a car wash. Since a fly sheet will air-dry in 30 minutes or less on a summer day, it can be put back on a horse soon after washing.

16.4 WET THOROUGHLY
Spread the fly sheet out flat on a concrete pad and wet it thoroughly with warm water to begin loosening the dirt.

16.5 SCRUB WITH PUSH BROOM
Add detergent to a bucket of warm water. Pour some soapy water on the sheet and scrub it with a stiff push broom. Stand on the sheet to help hold it flat while scrubbing. Flip it over and scrub the other side.

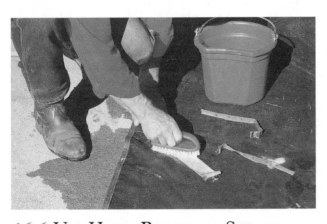

16.6 USE HAND BRUSH ON STRAPS
Lay the straps on top of the sheet and scrub both sides with a stiff brush. Pay particular attention to the trim at the shoulders and tail, where body oils concentrate. You may want to presoak these areas with a spot cleaner or a concentrated solution of your laundry detergent before you begin. If you use bleach, be aware it might change the color of some trim.

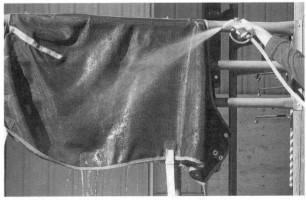

16.7 RINSE, RINSE, RINSE
Hang the sheet over a fence or rail and rinse thoroughly. Flip the sheet over and rinse again to remove every trace of detergent and bleach residue that could cause skin irritation.

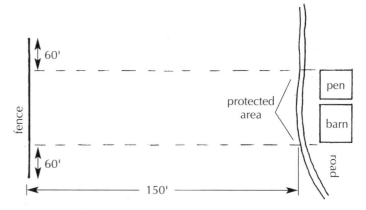

16.8 BLANKETS

Take winter blankets out of storage and check them over to make sure they'll be ready to put on a horse at a moment's notice. Check all straps, snaps, buckles, and elastic. This should have been done in the spring when the blankets were stored, but give a double check now.

SNOW FENCE LAYOUT

Place the snow fence about 150 feet (46 m) from the area you want to protect, such as the road and buildings, and perpendicular to the prevailing winter wind. Extend the fence 60 feet (18.3 m) past both ends of the protected area. Traditional snow fence of red wooden pickets is half the price of plastic snow fence and just as effective.

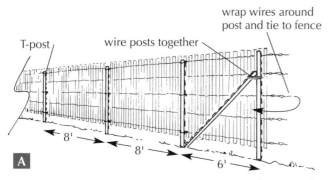

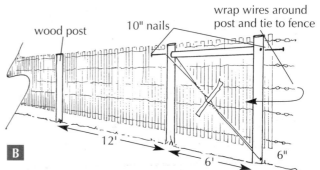

SNOW FENCE INSTALLATION

Attach the snow fence to 6-foot (1.8-m) T-posts spaced 8 feet (2.4 m) apart (A), or to 5-inch by

8-foot (12.7-cm by 2.4-m) wood posts spaced 12 feet (3.7 m) apart (B). Leave a 6-inch (15.2-cm) gap between the fence and the ground.

Snow Shovels

Unlike snowblowers, shovels are inexpensive, maneuverable in tight places, and will start working as soon as you do. Wide, short snow shovels work well for pushing an inch or two of snow off a hard surface like a concrete walk, but they're not well suited for lifting and moving deep snow. When you need to clear a deep drift, it's hard to beat a scoop shovel. Because of their light weight, plastic and aluminum scoop shovels are a good choice for digging through deep snow or drifts. A steel shovel is

harder, heavier, and more effective at removing hard-packed snow and will stand up to chipping ice without bending or cracking.

Rock salt or a snowmelt product can be used to melt layers of ice, but be aware that accumulated salt can kill grass and other vegetation and prevent it from growing for some time. Sand, gravel, sawdust, and cat litter can provide immediate traction on ice or hard-packed snow. Of these, sand and gravel work best. Sawdust is an insulator and will prevent the snow and ice underneath from thawing. Cat litter can get slimy and slippery as it soaks up moisture.

Insulating Pipes

Heat cable is taped to the bottom of a plastic or metal pipe (A) or wrapped around it (B) to ensure the pipe doesn't freeze and burst. The cable plugs into a standard outlet and has a built-in thermostat that turns on at around 35°F (1.6°C) and off at around 45°F (7.2°C).

The cable's thermostat is positioned at the coldest end of the pipe. Wrapping plastic pipes with aluminum foil first will provide more even heat distribution. Never operate a heat cable on a plastic pipe that doesn't contain water, and don't use heat cable on a garden hose.

If insulation is used, the cable goes on first: the cable thermostat should contact the pipe and be covered with insulation. Closed-cell polyethylene foam insulation comes in the form of a black tube 3 to 6 feet (0.9 to 1.8 m) long that's split lengthwise, making it easy to slip over a pipe (C). Fiberglass insulation up to ½ inch (12.7-mm) thick and 3 inches (7.6 cm) wide comes in rolls and is wrapped around the pipe in a spiral (D).

Be sure to cover all pipe fittings and joints with insulation. This is one case in which more is not better. Using too much insulation (more than ½ inch [12.7 mm] of fiberglass) may cause the cable to overheat, causing fire. The cable should be unplugged when the outside temperature is above 50°F (10°C) and checked periodically for rodent damage.

Study instructions. Some cables cannot be used within 1 foot (0.3 m) of combustible materials, which rules out using them in most walls, floors, and ceilings.

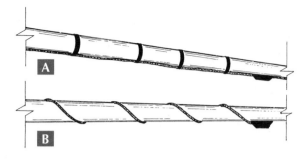

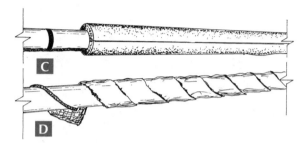

PIPE INSULATION

A. Heat cable taped to bottom of pipe. **B.** Heat cable wrapped around pipe. **C.** Foam tube insulation. **D.** Fiberglass wrap insulation.

Snow Pads

Ask your farrier to add snow pads to your horse's shoes. Tube-type rim pads, or "tubes," are the best way to prevent snow from balling up on the bottom of a shod horse's feet. Unlike full pads, tubes do not diminish a horse's traction.

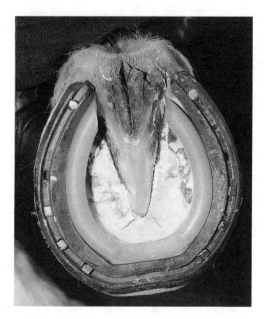

16.9 **Tube pads are used for winter riding.**

Winter

Checklist

❏ Keep up on hoof care during winter; even if horses are barefoot, have them trimmed every 6 to 8 weeks.

❏ Make sure water sources remain open so horses can drink (see chapter 13).

❏ When weather prevents riding, use time to clean and repair tack.

16.10 OPEN CREEKS AND PONDS

If the ice is too thick for horses to break, make a hole in the ice when horses are most likely to drink, such as after they've eaten their morning roughage. Ice fishermen use a heavy iron bar called a "spud" that has a chisel point to chop holes through several feet of ice. An ax will suffice if the ice is not too thick.

16.12 THERMOSTAT

Electric heating devices can be turned on automatically using this small thermostat (encased in white plastic) that plugs into an ordinary outlet (see Appendix). The thermostat turns on devices plugged into it when the temperature drops to 34°F (1.1°C) and turns them off when the temperature reaches 46°F (7.8°C). It can be used, for example, to turn on heat cables attached to pipes, a small heater near a medicine cabinet, or a light bulb or heat lamp, as shown here, that might be directed at water pipes or a faucet.

16.11 MAKE A DRINKING HOLE IN TROUGH

A Klim-Bonker (see Appendix), an ice-breaking tool designed for horsemen, makes an 8-inch (20.3 cm) hole in the ice, which is the perfect size for a horse to drink from. The remaining ice helps insulate the water.

Cleaning and Conditioning Tack

Winter is the perfect time to give tack a thorough cleaning and conditioning. To protect leather tack, clean it, feed it, and seal it (photo 16.13). Use a soft brush and a dry towel to remove dust, then wipe off the surface dirt with warm water. Next, use saddle soap or a leather cleaner to lift the dirt out of the leather (see appendix). A toothbrush works well for cleaning tooling. Remove residue and allow the leather to partially dry. Treat the flesh side (coarser underside) of the leather with a leather conditioner. Once the conditioner has soaked into the leather and dried, seal the leather. Applying a leather finisher or a coat of glycerin saddle soap will make it easier to keep clean. Let it dry for 30 minutes, and buff it with a chamois.

Never use linseed oil or mineral oil because it hardens leather. Don't soak tack in oil, as it can weaken the stitching and cause the leather to stretch.

16.13 Use those stormy winter days to give your tack a thorough cleaning.

Spring

Checklist

- ❏ Vaccinate.
- ❏ Arrange for dental check.
- ❏ Wash winter blankets and store them away.
- ❏ Shoe barefoot horses going back into work.

Medical and Dental Care

Most horses need annual vaccinations for tetanus, encephalomyelitis (sleeping sickness), influenza, and rhinopneumonitis. Ask your vet if your horses need any other vaccinations. Early spring is the traditional time for annual vaccinations because it allows enough time for the antibody titer in the blood to increase before insect season. Insects are carriers of several equine diseases, such as sleeping sickness. This timing also ensures horses are protected against influenza outbreaks, which tend to occur in cold, wet weather of spring and fall and when horses congregate at shows or other group activities. Scheduling your horse's annual vaccinations at the same time each year will help keep your horses on track.

16.14 An annual visit from a vet or equine dentist is required to "float," or file, the sharp points and edges of a horse's teeth that could cut the tongue or inside of the cheeks. Young horses also should be checked for retained caps (baby teeth), and for the presence of wolf teeth, which may need to be removed.

Washing Winter Blankets

Horses love rolling in mud, especially in spring, when shedding hair makes them itch. Instead of throwing winter blankets on a heap where they can mold and be ruined, now is the time to wash winter blankets and store them until next winter. Begin by brushing off as much dirt as you can while the blanket is still on the horse.

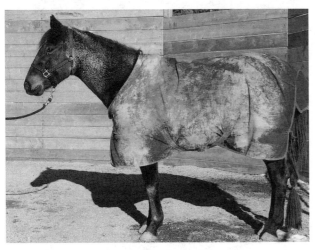

16.15 **Horses love to roll in spring mud.**

Preparing a Blanket for the Wash

1. The hook portion of hook-and-loop closures, on winter blankets especially, fills with horse hair and needs frequent cleaning to keep its grip. First, lift the hair pads using a large needle.
2. Make a loop of duct tape and press it onto the hook-and-loop to remove more hair.
3. Use a small brush to remove the last of the hairs. Then seal (see Laundry Tips).
4. If the blanket has wither fleece, lay the blanket on a flat surface and use a stiff brush to clean and fluff the fleece so it washes better.
5. Brush or vacuum the inside of the blanket to remove as much hair as you can. Otherwise, the hair can clog up the washer and cover the entire blanket with hair.
6. Untie tail cords; unbuckle leg, chest, and surcingle straps.
7. If the outer shell of the blanket is waterproof, put the blanket in the washing machine inside out. If it's not waterproof put it in with the dirty side out. Many blankets are too large to wash in home-size top-load machines and some are damaged by the twisting forces of a center agitator.

LAUNDRY TIPS

▶ Many laundromats prohibit washing horse blankets because the hair is damaging to the machines and the smell drives customers away. Preserve good relations with your local laundromat by precleaning your blankets to remove as much hair as possible before bringing them in. When you're through at the laundromant, either wipe all hair from the inside of the machine or run the empty washing machine through a cycle to clean it out.

▶ Buy extra "loop" fabric and cut it to cover the clean "hook" portions of fasteners. This will keep the hooks clean while the blanket is laundered.

▶ To keep buckles from banging and damaging the inside of the washing machine, remove as many straps and buckles as possible and wash separately. For those buckles that can't be removed, cover each with a sock secured by a rubber band.

▶ A large-capacity front-load machine is best for large winter blankets because there is no agitator to tangle the straps and wring the waterproof coating. The high volume of water that flushes through the blanket results in superior washing and rinsing. Use a minimum amount of mild detergent or soap.

Drying and Storage

Most winter blankets must be drip-dried. It takes 1 to 3 days for a blanket to dry thoroughly. The blanket should be turned inside out several times as it dries. Some blankets are designed to be dried in a dryer.

When blankets are completely dry, put them in bags with mothballs or cedar chips to prevent insect damage, and store them in a trunk or on a shelf in a rodent- and insect-free place (see photo 16.8).

16.16 BLANKET DONUTS
To keep surcingle buckles from coming undone, slip a rubber donut over the tongue portion of the buckle. Rubber donuts (top) are sold in tack shops and catalogs, or you can use castrating rings (bottom), available from a vet or feed store.

Summer

Checklist

- ❑ Spray weeds with a horse-safe herbicide.
- ❑ Keep grass mowed around stable and pens.
- ❑ Repair and paint fences and other facilities.
- ❑ Clean hay storage area and fill with new hay.
- ❑ Train and ride, train and ride.

16.17 MOWER MAN
Keeping grass and weeds trimmed at least 20 feet (6.1 m) from buildings and pens reduces fire danger, rodent and tick infestation, and fly-breeding sites. Also, trimming around pens can discourage horses from pushing through the pens to graze.

16.18 LAUGHING LADY
Remember what all this stablekeeping is for: to enjoy your horses!

APPENDIX

Resources

County Agricultural Agent: To find your county extension office, check the government section of your phone book under County Government for Cooperative Extension Office.

Supplies and Materials: For supplies and materials mentioned in this book, first look in your local tack shop or farm store. If you are unable to find what you need, call the following manufacturers to find the dealer nearest you, or contact one of the mail-order companies listed under Muzzles below.

Antichew Products

Carbolineum Wood Preserving Co.
P.O. Box 090348
Milwaukee, WI 53209
(800) 671-0093
Halt Cribbing liquid

**Classic Equine
Equipment, Inc.**
Rt. 2 Box 681
Ironton, MO 63650
(800) 444-7430
www.classic-equine.com/
Metal no-chew edging

Cowboy Center
3221 NW 79th Street
Miami, FL 32147
(800) 691-4122
www.cowboycenter.com/
Cribox paste

Dyco-Tec Products, Ltd.
29 W. 602 Schick Road
Bartlett, IL 60103
(630) 837-6410
Dyco-Sote liquid

Farnam Companies, Inc.
P.O. Box 34820
Phoenix, AZ 85067-4820
(800) 548-2828
www.farnam.com/
Chew Stop, Cayenne Hot Spray, and No Chew liquids

Miller's Harness Company
235 Murray Hill Parkway
East Rutherford, NJ 07073
(800) 553-7655
Cribox paste

Northern Light Stalls
1438 County Road G
New Richmond, WI 54017
(800) 246-3190
www.nlstalls.com/
Metal no-chew edging

Carts

Homestead Carts
955 E. Ellendale Avenue
Dallas, OR 97338
(503) 623-6423
Homestead carts

Rubbermaid Specialty Products
1147 Akron Road
Wooster, OH 44691
(330) 264-6464
www.rubbermaid.com/
Farm and garden carts

True Engineering, Inc.
999 Roosevelt Trail
Windham, ME 04062
(800) 366-6026
www.smartcarts.com/
Smart Cart

Fly Controls

Big D Products
490 Watt Drive
Fairfield, CA 94585
(800) 864-2443
Fly sheets, blankets

Cashel Company
446 Gore Road
Onalaska, WA 98570
(800) 333-2202
Fly masks

The Curvon Corporation
34 Apple Street
Tinton Falls, NJ 07724
(800) 631-2236
Fly sheets, blankets

Custom-Made Saddlery
151 N. Reservoir Street
Pomona, CA 91767
(909) 469-1240
Fly sheets

Farnam Companies, Inc.
P.O. Box 34820
Phoenix, AZ 85067-4820
(800) 548-2828
www.farnam.com/
*Fly sheets, fly masks, Swat cream,
Equitrol feed-through larvicide*

Horseware Triple Crown Blanket
P.O. Box 6328
Kinston, NC 28501-0328
(800) 887-6688
www.horseware.com/
Fly sheets, turnout sheets, and blankets

Loveland Industries
P.O. Box 1289
Greeley, CO 80632
(800) 356-7202
Sticky Fly Tape System

Neogen Corporation
628 Winchester Road
Lexington, KY 40505
(800) 477-8201
www.neogen.com/animalsafe.htm
Spray and cream repellent

Royal Riders
120-A Mast Street
Morgan Hill, CA 95037
(800) 437-6676
www.royalriders.com/
Fly sheets

Spalding Laboratories
760 Printz Road
Arroyo Grande, CA 93420
(800) 845-2847
www.spalding-labs.com/
Biological fly control

Muzzles

Dover Saddlery
P.O. 5837
Holliston, MA 01746
(800) 989-1500

Horse Health USA
P.O. Box 9101
Canton, OH 44711-9101
(800) 321-0235
www.horsehealthusa.com/

Jeffers Equine
P.O. Box 948
West Plains, MO 65775-0948
(800) 533-3377

Libertyville Saddle Shop
P.O. Box M
Libertyville, IL 60048-4913
(800) 872-3353
www.saddleshop.com/

Miller's Harness Company
235 Murray Hill Parkway
East Rutherford, NJ 07073
(800) 553-7655

State Line Tack
P.O. Box 935
Brockport, NY 14420-0935
(800) 228-9208
www.statelinetack.com/

United Vet Equine
14101 W. 62nd Street
Eden Prairie, MN 55346
(800) 328-6652
www.unitedvetequine.com/

Valley Vet Supply
P.O. Box 504
Marysville, KS 66508-0504
(800) 356-1005
www.valleyvet.com/

Weise Equine Supply
P.O. Box 920
Brockport, NY 14420-0920
(800) 869-4373

Odor Control Products

Advanced Environmental Solutions
4214 E. Indian School Road,
   Suite 201A
Phoenix, AZ 85018
(888) 921-4590
www.aesbio.com/

**Church & Dwight
Company, Inc.**
469 N. Harrison Street
Princeton, NJ 08543
(800) 526-3563
www.ahdairy.com/products/

G.G. Bean, Inc.
P.O. Drawer 638
Brunswick, ME 04011-0638
(800) 238-1915
www.ggbean.com/

Steelhead Specialty Minerals, Inc.
N. 1212 Washington, Suite 12
Spokane, WA 99201
(800) 367-1534
www.s-s-m.com/

Westhawk Traders
1810 Alberni Street, Suite 203
Vancouver, BC CANADA
V6G 1B3
(800) 663-1241

Psyllium

Farnam Companies, Inc.
P.O. Box 34820
Phoenix, AZ 85067-4820
(800) 548-2828
www.farnam.com/
Sand Clear 99

Gateway Products
P.O. Box 529
Holly, CO 81047
(888) 474-6367
www.gatewayproductsinc.com/
Su-Per Psyllium

Rubber Mats

Note: Call for dealer nearest you, as shipping adds a significant cost to these heavy items.

Caple-Shaw Industries, Inc.
1112 NE 29th Street
Fort Worth, TX 76106
(800) 969-3234
www.capleshaw.com/
Black Beauty Mats

Humane Manufacturing Company
805 Moore Street
Baraboo, WI 53913-2796
(800) 369-6263
www.humanemfg.com/
Lok Tuff Mats

Linear Rubber Products, Inc.
5416 46th Street
Kenosha, WI 53144
(800) 558-4040
www.rubbermats.com/ssm.html

North West Rubber Mats Ltd.
33850 Industrial Avenue
Abbotsford, BC CANADA
V2S 7T9
(800) 663-8724
Red Barn Mats

RB Rubber Products, Inc.
904 E. 10th Avenue
McMinnville, OR 97128
(800) 525-5530
www.rbrubber.com/
Tenderfoot Mats

Summit Flexible Products Ltd.
P.O. Box 520
Buckner, KY 40010
(800) 782-5628
www.summitflex.com/
Protector Mats, Mighty Lite Mats, rubber bricks, and tiles

Stall Forks, Plastic

Farnam Companies, Inc.
P.O. Box 34820
Phoenix, AZ 85067-4820
(800) 548-2828
www.farnam.com/

K & D Plastics, LP
4430 West Hwy 82
Gainesville, TX 76240
(800) 635-7919
www.kdplastics.com/

Vacuums

Electric Cleaner Company, Inc.
P.O. Box 400
Osseo, WI 54758
(800) 456-9821
www.electriccleaner.com/
Electro-Groom

Metropolitan Vacuum Cleaner
P.O. Box 149
Suffern, NY 10901
(800) 822-1602
www.metrovacworld.com/
Vac-n-Blow

Miscellaneous

Aquatic Control
P.O. Box 100
Seymour, IN 47274
(800) 753-5253
ag.ansc.purdue.edu/aquacon/
Stocktrine II algae treatment

Co-Line Welding, Inc.
1232 100th Street
Sully, Iowa 50251
(800) 373-7761
Sure Latch gate latches

Colorado Kiwi Company
1821 Kamar Plaza #4
Steamboat Springs, CO 80487
(800) 345-8846
Kiwi latches

EasyHeat, Inc.
31977 US 20 East
New Carlisle, IN 46552
(800) 562-6587
www.easyheat.com/
Heat cable, thermostats, and heat mats

Elite Equestrian Products
9591 Sulphur Road
Sulphur, KY 40070
(800) 544-5819
Floor magnets

Hamilton Products
P.O. Box 770069
Ocala, FL 34477-0069
(800) 521-9238
www.hamiltonproducts.com/
Halters

Johnson Barns & Trailers
22307 N. Black Canyon Hwy.
Phoenix, AZ 85027
(602) 465-9000
Custom barns

Kalglo Electronics Co., Inc.
5911 Colony Drive
Bethlehem, PA 18017-9348
(888) 452-5456
www.kalglo.com/products.htm
Infrared heaters

Richard Klimesh
P.O. Box 140
Livermore, CO 80536
(970) 221-2948
www.horsekeeping.com/
*Klim-Bonker ice-breaker tool,
blanket rods*

Leatherman Tool Group, Inc.
P.O. Box 20595
Portland, OR 97220-0595
(800) 847-8665
www.leatherman.com
Multipurpose tool

Leather Therapy
Unicorn Editions, Ltd.
P.O. Box 432
Oldwick, NJ 08858
(800) 711-8225
www.leathertherapy.com/
*Leather cleaning and conditioning
products*

Life Data Labs
P.O. Box 349
Cherokee, AL 35616
(800) 624-1873
www.lifedatalabs.com/
Farrier's Formula hoof supplement

Novi International
P.O. Box 8310
Scottsdale, AZ 85252
(800) 842-5378
www.keepsafer.com/
Intercom system

Priefert Manufacturing Company
P.O. Box 1540
Mt. Pleasant, TX 75455
(800) 527-8616
Panels and gates

Purina Mills, Inc.
P.O. Box 66812
St. Louis, MO 63144
(800) 227-8941
horse.purina-mills.com/
Feed

Steinbau Construction
RD.2, Box 158
Pleasant Valley, NY 12569
(914) 635-9606
www.castlemall.com/steinbau/
Custom barns

Wilsun Equestrian
2210 McFarland/400 Boulevard
Alpharetta, GA 30004
(800) 942-5567
www.wilsunblanket.com/
Blankets

Recommended Reading

Ewing, Rex A. *Beyond the Hay Days: A Refreshingly Simple Guide to Effective Horse Nutrition.* LaSalle, CO: PixyJack Press, 1997.

Fershtman, Julie I. *Equine Law and Horse Sense.* Franklin, MI: Horses and the Law Publishing, 1996.

Hayes, Karen. *Emergency! The Active Horseman's Book of Emergency Care.* Middletown, MD: Half Halt Press, 1995.

———. *Hands-On Horse Care from Horse & Rider.* North Pomfert, VT: Trafalgar Square Publishing, 1997.

Hill, Cherry. *Horse Handling and Grooming.* Pownal, VT: Storey Books, 1990.

———. *Horse Health Care.* Pownal, VT: Storey Books, 1997.

———. *Horsekeeping on a Small Acreage.* Pownal, VT: Storey Books, 1997.

———. *Trailering Your Horse: A Visual Guide to Safe Training and Traveling.* Pownal, VT: Storey Books, 2000.

———. *Your Pony, Your Horse.* Pownal, VT: Storey Books, 1995.

Hill, Cherry, and Richard Klimesh. *Maximum Hoof Power: How to Improve Your Horse's Performance Through Proper Hoof Management.* New York: Howell Book House, 1994.

Kellon, Eleanor. *Dr. Kellon's Guide to First Aid for Horses.* Ossining, NY: Breakthrough, 1990.

Lewis, Lon D. *Feeding and Care of the Horse,* 2nd ed. Media, PA: Williams & Wilkins, 1996.

✦ INDEX ✦

Note: Numbers in *italics* indicate an illustration; numbers in **boldface** indicate a chart.

Mats
  feeding on, 96–97
  for flooring, 5, **6–7,14,** 15
  installation, 16
Medical care (horse), 141
Metal edging to prevent chewing,
    62, 63
Mice, pest control, 78
Mineral supplements, 87, 90
Moldy grain, 88
Moldy hay, 81
Mud control, 49, 58, 110
Muzzles for vices, 62

Natural disasters, 123
Newsprint bedding, 64, 65, **66**
911 for emergencies, 121
Nondraining stall flooring, 15

Oats, 85
Odor Capture (deodorizer), **68**
Odor control products, 68, **68**
Outlet caps for safety, 118
Outlet covers, waterproof, 35
Oxyuris equi (pinworms), *71*

Pacing, **60**
Paddocks. *See* Turnout areas
Palpation chutes (stocks), 30
Panels, turnout areas, 50, 51, 52,
    55, *55*
Panic snaps, 26
Paper bedding, 64, 65, **66**
Parasite control
  for ascarids (roundworms), 71
  for bots (*Gasterophilus*), 71, 72,
    72
  colic, 71
  deworming program, 71, 135,
    136
  feed-through wormers, 71
  ivermectin, 71
  paste dewormers, 71
  for pinworms (*Oxyuris equi*),
    71
  for strongyles (bloodworms), 71
  thromboembolic colic, 71
Parasitoids (fly predators), 73

Paste dewormers, 71
Pasture hay feeders, 98
Pasture waterers, 104
Pawing, 49, 56, 59, **60**
Pelleted feeds, 86
Pens. *See* Turnout areas
Permanent pens, 55
Pest control, 71–78. *See also*
    Sanitation
  for birds, 78
  feed-through larvicide for flies,
    75
  for flies, 73–77
  for hornets, 78
  for mice, 78
  parasitoids, 73
  for rodents, 78
  stable pests, 78, *78*
  for wasps, 78
Phillips screwdrivers, 44, *44*
Phones for emergencies, 121
Phosphorus and grain, 85
Pinworms (*Oxyuris equi*), *71*
Pipes, insulating, 139, *139*
Pipes, thawing frozen, 133, *133*
Pliers, 44, *44*
Plumbing, 20, 116
Poisonous plants in hay, 80
Ponds, water from, 103, 140
Power loss, emergency, 123–24
Premise fly sprays, 76
Propane heaters, 126–27
Psyllium for sand colic, 88

Railroad ties, 56
Rakes, 45, *45*
Ration calculation, 91–94
Ready-to-use (RTU) fly sprays, 76
Records storage, 17
Refrigerator in tack room, 17
Repair areas, 22–23
Resurfacing, turnout areas, 57
Retriever truck, hay, 82
Ribs and horse's condition, 93
Road base flooring, **6–7**
Rodents, pest control, 78
Rolled grain, 85–86, 88
Ropes storage, 37, 42, *42*

Roughage. *See* Hay
Roundworms (ascarids), 71
RTU (ready-to-use) sprays, 76
Rubber
  connectors, 53
  flooring, **6–7**
  mats, feeding on, 96–97
  mats, flooring, 5, **6–7,14,** 15,
    16
Runs. *See* Turnout areas

Saddle blankets storage, 37, 41
Saddle storage, 17, 18
Safety, 113–19
  barns, 114, *114*, 116–19, *117*
  and dogs, 113
  facilities, 113, *115*, 115–17,
    117
  feeders, 116
  fences for, 115
  fire extinguishers, 116–117,
    129, 135
  fire prevention, 116–18, *117*
  first aid kit, 122
  flooring, 113, 119
  handling horses, 113
  hay, 117
  lighting for, 119
  plumbing, 116
  security for, 119
  smoking policy for, 118
  storage areas, 37
  tack, 113, 119
  tools, 113
  turnout areas, 49, 51, 52, 54
  watch animals for, 119
  water tanks, 107
  work areas, 24, 27
Salt (sodium chloride), 87
Sand colic, 14, 49, 64
Sand flooring, **6–7**
Sanitation, 64–70. *See also* Pest
    control; Stalls
  bedding, 64–66, **66**
  cleaning stalls, 66–67
  composting, 69
  disposal, 69
  for manure, 64, 67, 69–70

Other Storey Titles You Will Enjoy

Building Small Barns, Sheds & Shelters, by Monte Burch. Covers tools, materials, foundations, framing, sheathing, wiring, plumbing, and finish work for barns, woodsheds, garages, fencing, and animal housing. 248 pages. Paperback. ISBN 0-88266-245-7.

The Horse Behavior Problem Solver, by Jessica Jahiel. Using a friendly question-and-answer format and drawing on real-life case studies, Jahiel explains how a horse thinks and learns, why it acts the way it does, and how you should respond. 352 pages. Paperback. ISBN 1-58017-524-4.

Horse Handling & Grooming: A Step-by-Step Photographic Guide, by Cherry Hill. This user-friendly guide to essential skills includes feeding, haltering, tying, grooming, clipping, bathing, braiding, and blanketing. The wealth of practical advice offered is thorough enough for beginners, yet useful enough for experienced riders improving or expanding their skills. 160 pages. Paperback. ISBN 0-88266-956-7.

Horsekeeping on a Small Acreage, by Cherry Hill. Thoroughly updated, full-color edition of the best-selling classic details the essentials for designing safe and functional facilities whether on one acre or one hundred. Hill describes the entire process: layout design, barn construction, feed storage, fencing, equipment selection, and much more. 320 pages. Paperback. ISBN 1-58017-535-X.

Stable Smarts, by Heather Smith Thomas. Gathered here in a readily accessible handbook, Thomas's hundreds of useful tidbits—gleaned over a lifetime of working with horses day in and day out—will generally improve and simplify the quality of life on a horse farm, while saving time and money in ways you never thought possible. 320 pages. Paperback. ISBN 1-58017-610-0.

Storey's Guide to Raising Horses, by Heather Smith Thomas. Whether you are an experienced horse handler or are planning to own your first horse, this complete guide to intelligent horsekeeping covers all aspects of keeping a horse fit and healthy in body and spirit. 512 pages. Paperback. ISBN 1-58017-127-3.

Trailering Your Horse: A Visual Guide to Safe Training and Traveling, by Cherry Hill. An essential guide to safe training and traveling, with detailed information on selecting an appropriate truck and trailer, training a horse to load and unload, and helpful traveling advice. 160 pages. Paperback. ISBN 1-58017-176-1.

These and other books from Storey Publishing are available wherever quality books are sold or by calling 1-800-441-5700. Visit us at www.storey.com.